D1452538

THEOLOGY AND SCIENCE AT THE FRONTIERS OF KNOWLEDGE

NUMBER FOUR

TRADITION AND AUTHORITY IN SCIENCE AND THEOLOGY

THEOLOGY AND SCIENCE AT THE FRONTIERS OF KNOWLEDGE

1: T. F. Torrance, *Reality and Scientific Theology*.
2: H. B. Nebelsick, *Circles of God, Renaissance, Reformation and the Rise of Science*.
3: Iain Paul, *Science and Theology in Einstein's Perspective*.
4: Alexander Thomson, *Tradition and Authority in Science and Theology*.
5: R. G. Mitchell, *Einstein and Christ, A New Approach to the Defence of the Christian Religion*.
6: W. G. Pollard, *Transcendence and Providence, Reflections of a Physicist and Priest*.

THEOLOGY AND SCIENCE AT THE FRONTIERS OF KNOWLEDGE

GENERAL EDITOR — T. F. TORRANCE

TRADITION AND AUTHORITY IN SCIENCE AND THEOLOGY

with reference to the thought of Michael Polanyi

ALEXANDER THOMSON

SCOTTISH ACADEMIC PRESS
EDINBURGH
1987

Published in association with the
Center of Theological Enquiry
and
The Templeton Foundation
by
SCOTTISH ACADEMIC PRESS
33 Montgomery Street, Edinburgh EH7 5JX

First published 1987

ISBN 0 7073 0452 0

British Library Cataloguing in Publication Data

Thomson, Alexander
Tradition and authority in science and technology:
with reference to the thought of Michael Polanyi.
1. Religion and science
I. Title
261.5'5 BL240.2

ISBN 0 7073 0452 0

GENERAL FOREWORD

A VAST shift in the perspective of human knowledge is taking place, as a unified view of the one created world presses for realisation in our understanding. The destructive dualisms and abstractions which have disintegrated form and fragmented culture are being replaced by unitary approaches to reality in which thought and experience are wedded together in every field of scientific inquiry and in every area of human life and culture. There now opens up a dynamic, open-structured universe, in which the human spirit is being liberated from its captivity in closed deterministic systems of cause and effect, and a correspondingly free and open-structured society is struggling to emerge.

The universe that is steadily being disclosed to our various sciences is found to be characterised throughout time and space by an ascending gradient of meaning in richer and higher forms of order. Instead of levels of existence and reality being explained reductionistically from below in materialistic and mechanistic terms, the lower levels are found to be explained in terms of higher, invisible, intangible levels of reality. In this perspective the divisive splits become healed, constructive syntheses emerge, being and doing become conjoined, an integration of form takes place in the sciences and the arts, the natural and the spiritual dimensions overlap, while knowledge of God and of his creation go hand in hand and bear constructively on one another.

We must now reckon with a revolutionary change in the generation of fundamental ideas. Today it is no longer philosophy but the physical and natural sciences which set the pace in human culture through their astonishing revelation of the rational structures that pervade and underlie all created reality. At the same time, as our science presses its inquiries to the very boundaries of being, in macrophysical and microphysical dimensions alike, there is being brought to light a hidden traffic between theological and scientific ideas of the most far-reaching significance for both theology and science. It is in that situation where theology and science are found to have deep mutual relations, and increasingly cry out for each other, that our authors have been working.

The different volumes in this series are intended to be geared into this fundamental change in the foundations of knowledge. They do not present "hack" accounts of scientific trends or theological fashions, but are intended to offer inter-disciplinary and creative interpretations which will themselves share in and carry forward the new synthesis transcending the gulf in popular understanding between faith and reason, religion and life, theology and science. Of special concern is the mutual modification and cross-fertilisation between natural and theological science, and the creative integration of all human thought and culture within the universe of space and time.

What is ultimately envisaged is a reconstruction of the very foundations of modern thought and culture, similar to that which took place in the early centuries of the Christian era, when the unitary outlook of Judaeo-Christian thought transformed that of the ancient world, and made possible the eventual rise of modern empirico-theoretic science. The various books in this series are written by scientists and by theologians, and by some who are both scientists and theologians. While they differ in training, outlook, religious persuasion, and nationality, they are all passionately committed to the struggle for a unified understanding of the one created universe and the healing of our split culture. Many difficult questions are explored and discussed, and the ground needs to be cleared of often deep-rooted misconceptions, but the results are designed to be presented without technical detail or complex argumentation, so that they can have their full measure of impact upon the contemporary world.

The author of this volume, the Rev. Dr Alexander Thomson, is both a biochemist and a theologian who has been deeply influenced by the writings of Michael Polanyi in which scientists and philosophers alike are challenged to rethink the received concept of knowledge in the light of the actual ways in which scientific discoveries are made. In this particular work the author is concerned to explore the deep affinity between natural science and Christian theology in respect of their commitment to the authority of the truth within appropriate frameworks of knowledge and institutional structures as they have developed in processes of inquiry. Dr Thomson makes considerable use of Michael Polanyi's distinction between subsidiary and focal awareness, implicit and explicit, or informal and formal thought (evident in the acquisition of scientific skills and in the scientific discovery of new truth), to examine the nature of understanding, authority and interpretation in the Church. This is then applied to the field of Christian communication and education in a fresh and illuminating way which has been put to the test in Alexander Thomson's work as a parish minister in Scotland. Here we are shown how questions thrown out by natural scientific inquiry can point beyond themselves and give rise to fruitful developments in other fields of inquiry, where there is the same respect for the claims of objectivity and the same commitment to the authority of the truth over which we have no control, but in the service of which there is authentic freedom.

Thomas F. Torrance

Edinburgh,
Advent, 1984

CONTENTS

PREFACE

THIS book rests on research in which I was earlier engaged in the University of Edinburgh. It deals with the important role played by tradition and authority in the scientific community. That the scientific community operates within its own tradition and exercises authority may come as a surprise to many. It is still often assumed that science demands freedom from authority in order that the truth may be known. The instilling of beliefs and values in us, perhaps from childhood, within a certain tradition, is seen as a form of indoctrination. Such traditional indoctrination, it is claimed, must be resisted. So science has often been seen as the opponent of authority and tradition.

However, with the re-examination of scientific objectivity and the nature of scientific discovery carried through by Michael Polanyi such a view of the relation of science to authority is corrected. From a study of the part played by authority and tradition in actual on-going scientific activity, we can learn lessons relevant for our understanding of the role of authority and tradition within the Church. This may even have significant feed-back for our understanding and appreciation of the way in which authority and tradition function in the scientific enterprise.

I would like to thank Professor Thomas F. Torrance for all the guidance and help he has given me in writing this work, and also for his introducing me to the thought and writings of Michael Polanyi. I am also indebted to my wife who has given me nothing but encouragement and support in my studies, which in a busy parish, have meant that even less time has been spent with her and the family than would otherwise have been the case.

Alexander Thomson
Rutherglen, Glasgow

CHAPTER I

THE ROLE OF TRADITION AND AUTHORITY IN MICHAEL POLANYI'S THOUGHT

1. *A False Conception of Science*

FOR the past three centuries a materialistic conception of science has exercised a pervasive influence on all our thinking. It is still firmly entrenched in present day popular beliefs about science. This is evident in the fact that many people today regard science as establishing a knowledge that is self-evident, positive, demonstrable, unambiguous and "objective". A knowledge that is based exclusively on the data of experience, on "facts". It is still widely supposed that from these data of recorded observations and of experimental measurements, scientific propositions and the mathematical laws of nature can be deduced in a truly impersonal and detached way by a set of explicit rules. Mathematical functions, $x = f(y)$, expressing natural laws are assumed to be derived exclusively from measured data. Such a conception of science necessarily involves the rejection of anything that cannot be proved or disproved by experiment. It identifies truth only with that which can be impersonally demonstrated. As a consequence all personal beliefs have to be eliminated. They have to be discarded in order to achieve a proper scientific detachment. They cannot lay any claim to truth. This has obviously had grave consequences for religious and moral truth. Religious and moral beliefs are regarded as being purely subjective, a form of personal bias. As they cannot be impersonally demonstrated they are assumed to have no bearing on reality. Where they are in fact retained, as for example in the case of moral values, it is only on the basis of "detached explanatory principles".[1] These are supplied by Utilitarianism and Pragmatism.

The rise of this materialistic conception of science can be traced back to certain elements of Greek thought.[2] Copernicus and Kepler both stand in the Pythagorean tradition of Greek thought. Within that tradition nature was interpreted in terms of number and geometrical form. Number was the ultimate form of things and inherent in nature. This conception of science required the personal contemplation and recognition of the harmony of number and nature as well as a personal search for and appreciation of geometrical excellence. The feelings of elation which Polanyi points to in the works of Kepler[3] were not the psychological by-product of the act of discovery. Nor are they the sign of subjectivity. They are the sign that there exists in the person of the scientist an intellectual passion which appreciates the value, worth and importance of any discovery in science. This recognition of the role played by the knowing person in knowledge, which is seen clearly in the works of Kepler and Copernicus, is something we will return to consider a little further on.

With Galileo however a change begins to take place, a change that lies behind the present popular misconception of science. While still accepting the Pythagorean tradition that the world is perfectly ordered, Galileo embraces another line of Greek thought stemming from Democritus. Within this tradition of Greek thought the real nature of things consisted not in numbers but of matter in motion. This was a mechanical view of the world. Shape, size, motion, quantity, became the primary qualities of things, their other properties, for example colour, were derived from these. The motion of material particles underlay all phenomena. Number became a mere measurement of events. Mathematical theory became an instrument for expressing the mechanical motions of particles. Recorded observations became more and more important. So there began with Galileo the gradual eclipse of personal involvement in knowing, an eclipse that was even more firmly established by the success of Newton. All matter was assumed to be under the rule of mechanical laws. Science had become the study of things capable of measurement and calculation. There was less room for a personal appreciation of the importance and value of the facts of science. The impersonal ideal of knowledge began to be established.

The critical philosophy of Descartes and Locke contributed to the eclipse of the role of the knowing person in our understanding of knowledge, and further led our understanding of science along the road of impersonal detachment. John Locke maintained that all true knowledge is based on the evidence of the senses and he drew a distinction between faith and demonstrable knowledge. Knowledge must be derived from the objective facts by reason alone, from the measured data of experience by definite rules, explicitly laid down. While it was in the context of attempting to gain religious tolerance that Locke proposed his argument of doubt, to prevent the imposition of beliefs that are not demonstrable, nonetheless, the logical consequences of his doctrine were soon worked out. All human beliefs accepted on the grounds of authority or tradition were regarded as being purely subjective and had to be discarded in order to achieve a proper scientific detachment and a truly objective approach to the world. Faith and belief fell short of knowledge in that they fell short of empirical and rational demonstration.

This materialistic conception of science became unconditionally accepted. The standard of truth became that which could be objectively demonstrated. "For modern man has set up as the ideal of knowledge the conception of natural science as a set of statements which are 'objective' in the sense that their substance is entirely determined by observation."[4]

2. *The Road to the Recovery of a True Understanding of Natural Science*

Polanyi's philosophy is one that calls for the reform of this particular ideal of scientific objectivity. He develops what he calls a "Post-Critical Philosophy", a title that aptly describes his aim. He not only rejects, as many others have done, this false understanding of scientific objectivity that has dominated our thinking for so long, but he develops, and has been the first to develop, a comprehensive alternative to the detached impersonal approach that has been characteristic of our understanding of science. Indeed Richard Gelwick indicates the significance of Polanyi's work by suggesting that we describe it as a new paradigm.[5] He calls this new paradigm a "heuristic philosophy", as it has been Polanyi's examination of the way in which scientific discoveries are made that has led to his reform of our conception of knowledge. The road to the recovery of a true understanding of objectivity lies in this direction — an examination of experimental science as it is in fact practised by scientists.

Northrop has also pointed out that this false ideal of science, which we have looked at, has arisen because scientists are rightly concentrating on the subject matter of their science and pay little attention to the actual methods they use in obtaining their results.[6] He further points out that it was one of Einstein's contributions to the philosophy of science that he too developed an epistemology by an examination of the actual method by which scientific theory is arrived at, and not as positivism has done, by making scientific procedure fit their epistemological assumptions. When we examine the scientific process we find that there are two components to our scientific concepts; an empirical component and a theoretical component. As Northrop says: "The full meaning of verified mathematical physics is only given in part empirically in sense awareness or in denotatively given operations or experiments, and hence involves also meaning which only the imagination can envisage and which only deductively formulated, systematic, mathematical constructions, intellectually conceived, rather than merely sensuously immediate, can clearly designate".[7]

Scientific theories are the "creations", the inventions of the human mind. They are intellectually known and not logically derivable from empirical observations, and cannot therefore be reduced to what is purely empirical. Although experience and empirical data may suggest what our scientific theories should be, the latter cannot be deduced from the former. There is a "logical gap" between scientific concepts and empirically given experience. A scientist using his creative imagination hits upon the basic scientific concepts and "there is no method capable of being learned and systematically applied that leads to this goal".[8] These scientific concepts are not, as they may at first seem, purely subjective constructions in the mind of the scientist. They reveal and derive from the actual rational structure of the real world; they are formed under the impact which the real world makes upon our minds as we seek to understand it, and reflect that rationality. Einstein himself felt that nothing could be said about the way in which a scientist "creates" his theories and bridges the logical gap between theory and experience, but this is the area that Polanyi develops further.[9]

To substantiate this claim, that a study of the scientific process would show the ideal of a detached impersonal objectivity to be false, we look at three main strands of Polyani's thought.

First of all, Polanyi recalls how it was the Marxist challenge that first prompted him to examine more closely the nature and justification of science.[10] After the Revolution the Soviet authorities embarked on a policy of enforcing a Marxist philosophy on all branches of Science. In genetics for example the Mendelian laws of inheritance were rejected because they contradicted the Marxist philosophy of dialectical materialism. Here emerged a fundamental difference in belief in the nature of science between Marxist scientists in the East and non-Marxist scientists in the West. Parts of science carrying conviction in the West failed even to gain recognition in Soviet Russia. This could only be accounted for by our belief or disbelief in the new premises of science as proposed by the Marxists. In other words it became clear that every true description of science demands that account be taken of certain "scientific beliefs" about the nature of things — in total contrast to the teachings of empiricism, positivism, and the prevalent popular conception of science today. In answer to the question as to what forms the basis of our acceptance of science we cannot say "because it presents truth that is self-evident". We can only say in reply "because I believe so".[11]

When we move on secondly to examine the nature of experimental science itself we find further evidence refuting the false ideal of objectivity taught by empiricist and positivist interpretations of science. Jaki speaks of the leap that Newton had to take from his experimental data in order to formulate his laws of gravitation, "a leap which Newton practiced in postulating a perfectly exact physical world on the basis of data *not* in perfect agreement with the laws he formulated".[12] Here we again come up against the "logical gap" mentioned above. Newton's laws of gravitation could not be verified experimentally with perfect rigour. "The vision of the world it embodied was ultimately a creation of the mind, a leap from sensory data far beyond the range of the senses. But because that vision was rooted in data provided by nature, the vision could become a vigorous science".[13] There was no exact experimental verification of Newton's laws. Verification required the same heuristic insight, the same vision as Newton himself possessed. Polanyi makes a similar point when he says, "Even the most exact sciences must therefore rely on our personal confidence that we possess some degree of personal skill and personal judgement for establishing a valid correspondence with — or a real deviation from — the facts of experience".[14] The scientist has to acquire the skill of being able to know whether the discrepancies between theory and instrument readings are due to random errors, or represent a real deviation from the theory, or a refutation of the theory. There is no rule which he can impersonally apply which will tell him this. Therefore even the most exact sciences rely on the personal skill of the scientist, and this skill is a valid form of knowledge. Polanyi comments on this situation as follows,

"The slight gap between theory and instrument readings turns out to be thin only in the way the edge of a wedge is thin — a wedge that will prove thick enough at its base completely to separate "knowledge" from "detached objectivity". Personal, tacit assessments and evaluations, we see, are required at every step in the acquisition of knowledge — even "scientific knowledge".[15]

Involved in the making of these personal, tacit assessments and evaluations, are certain standards of accuracy, plausibility, intrinsic interest and profundity on which the scientist relies. Polanyi illustrates this by reference to the Velikovsky Affair and the gestation period in animals.[16] In spite of the very close agreement between theory and experiment data, in both these cases, they were nonetheless rejected by the scientific community as absurd. They were rejected because they were not plausible. This judgement of plausibility depends on our view of the nature of things and it profoundly affects the conclusions we draw from experimental data. In contrast to the Velikovsky Affair and the gestation period in animals, where there was good agreement between experimental data and theory, in the case of the periodic table it was not abandoned when certain elements could not be fitted into it in accordance with their atomic weights.[17] Such contradictions of the theory were regarded as "anomalies" and the theory of the periodicity of elements was not rejected.

There are therefore no strict rules for assessing the plausibility of any contribution to science. It is a personal assessment and therefore undemonstrable.

If we turn from this brief study of experimental science itself and look thirdly at the specific contributions made by Copernicus and Einstein to scientific knowledge we find further evidence calling for the rejection of the ideal of a "detached objectivity". We have already noted the fact that both Copernicus and Kepler stand in the Pythagorean tradition of Greek thought, and that the feelings of elation found in the works of Kepler were the sign that there exists in the person of the scientist an intellectual passion which appreciates the importance and the value of any discovery in science. Copernicus' theory itself was accepted because of the greater intellectual satisfaction which it gave, and which as such was regarded as being more objective. "Copernicus' heliocentric system was based upon the elegance and beauty of its explanations of the movements of the planets rather than upon the accumulation of new empirical observations".[18] Here our knowledge of the world relying on the immediate sensory experience of the motion of the sun in the sky and the apparent motionlessness of the earth was replaced by a knowledge relying more on theory and indeed contradicting immediate sensory experience. Scientific theories cannot therefore be "objective" in the sense that they are entirely determined by observation. It is rather that scientists have the capacity to intuit a rationality in nature and to create theories which reflect that rationality in nature. This is not as Polanyi points out, what we are taught today; "that the discovery of objective truth in science consists in the apprehension of a rationality which commands our respect and arouses our contemplative admiration; that such discovery, while using the experience of our senses as clues, transcends experience by embracing a vision of reality beyond the impression of our senses, a vision which speaks for itself in guiding us to an even deeper understanding of reality".[19] Such feelings of admiration, elation and satisfaction must be given an integral place in any account of the art of scientific discovery.

This is again seen with relativity. Relativity was accepted because of its intellectual beauty, "a beauty that exhilarates and a profundity that entrances us".[20] It accounted for previously known data in a more intellectually satisfying way and was accepted because of that. Not until a number of years later was there any experimental evidence to support it. Indeed it was accepted in the face of empirical evidence which appeared to contradict it.[21] Copernicus and Einstein would teach us therefore that such emotional appraisals of a scientific theory cannot be eliminated from any account of science and that the ideal of objectivity as fostered by empiricism and positivism which maintain that our scientific theories are held in a detached and pragmatic manner is false. Rather is it the case that we have the capacity to intuit a rationality in nature to which our theories direct us. We rely on our own personal judgement as to whether we continue to accept or reject our theories in the face of supporting or conflicting experimental evidence. Any description of science must therefore take account of the points brought out in the three areas surveyed above. It must take account of the beliefs with which the scientist works. Science does not progress by doubting and scepticism, but by a firm belief in discovering coherence and meaning in the world around us. Any attempt to understand science must take into account the scientist's skills and value judgements, and his emotional appraisal of a theory. Science does not give us a detached impersonal knowledge.

Polanyi gives us the following definition of scientific objectivity which incorporates all these points. He maintains "... that the discovery of objective truth consists in the apprehension of a rationality which commands our respect and arouses our contemplative admiration; that such discovery, while using the experience of our senses as clues, transcends experience by embracing the vision of a reality beyond the impression of our senses, a vision which speaks for itself in guiding us to an ever deeper understanding reality ...".[22]

We turn now to look in more detail at this alternative to the ideal of "detached objectivity" which Polanyi proposes, and to look further at how scientific theories are created and how the "logical gap" between theory and observation, problem and discovery is bridged.

3. Personal Knowledge — The Restoration of a Proper Understanding of Scientific Objectivity

In our previous section we noted that in the exact sciences the personal skill of the scientist comes into play as he tries to establish either a correspondence of instrument readings with their value as predicted by theory, or, a deviation of these readings from their predicted value. This skill is a form of knowledge and similar skills are found in other branches of science. This is evidenced by the fact that students of biology, medicine and chemistry spend a great deal of time in practical work learning the skills which will enable them to recognise the things that form the subject matter of their particular branch of science. "They are training their eyes, their ears, and their senses of touch to recognise *the things* to which their textbooks and theories refer. . . . Textbooks of chemistry, biology and medicine are so much empty talk in the absence of personal, tacit knowledge of their subject matter".[23] The medical student must acquire the skill of being able to diagnose a case of a certain illness correctly. This cannot be learned from a textbook alone — it is a personal skill acquired through practice and practical experience. Similarly a student of biology must learn the skill of being able to identify the species to which a plant belongs. This also is a skill, a knowledge that cannot be obtained from textbooks.

It is by an examination of the structure of such skills that Polanyi seeks to establish a proper understanding of the art of scientific discovery. He takes as his starting point an analysis of perception.

To be able to perceive an object is a skill. It is a skill that has to be learned as a child grows and which has to be developed by practice. As we look at things in front of us we have a desire to make out what it is that we are seeing, to pick out significant shapes, and make order and sense out of the multitude of visual clues presented to us. Our powers of perception have to be able to pick out very different appearances of the same object — we have to see it as an object of constant shape, size and colour even when it is moving before us.[24] Seeing an object as it is, is a skilful integration of a large number of clues, some of which cannot be experienced in themselves (adjustment of eye lens, memory etc.) and which Polanyi calls "subliminal clues"; and others which normally see in our field of vision which he calls "marginal" clues. It is a skilful integration of these clues, many of which are unspecifiable, into a meaningful whole, which enables us to see an object as it is. Furthermore in this act of integration we are not directly aware of either the subliminal or marginal clues. We attend from them to the object to whose appearance they contribute. Their isolated appearance as individual clues merges into the appearance of the whole.

A similar structure is found in all other skills, for example the recognition of a physiognomy, a skill like riding a bike or the skill of knowing how to use a tool. The striking feature in all of these skills is the presence of two different kinds of awareness. For example we use a tool by relying on it to accomplish something we intend. When using a hammer to drive in a nail, we are aware of the feeling of the hammer in our hands but our attention is directed away from this to the skilful act of driving in the nail, to this comprehensive achievement. We attend from the tool in our hands to this comprehensive achievement.

Polanyi devises a general terminology for the relation of a set of particulars to a comprehensive entity. "The essential features throughout is the fact that particulars can be noticed in two different ways. We can be aware of them uncomprehendingly, i.e., in themselves, or understandingly, in their participation in a comprehensive entity. In the first case we focus our attention on the isolated particulars; in the second, our attention is directed beyond them to the entity to which they contribute. In the first case therefore we may say that we are aware of the particulars focally; in the second, we notice them subsidiarily in terms of their participation in a whole".[25]

In perception then we are subsidiarily aware of many clues while seeing an object in front of us. If we looked at these visual clues in themselves they would lose all meaning. It is only as we attend from them to a focal awareness of the object, to whose appearance they contribute, that their meaning is established in our perception of this comprehensive entity. The clues in themselves are meaningless if we lose sight of the pattern, or appearance of the object which they jointly constitute.

Furthermore Polanyi points out that in all skills we

possess a knowledge which is not completely specifiable, "we know more than we can tell". We can skilfully use a tool without knowing how we actually use it. We may know how to ride a bicycle yet not be able to tell how we do it. In every comprehensive achievement and act of integration we merge together particulars many of which are not specifiable. This is true in the case of perception. Most of the clues which we use in seeing an object are never identified by us. In riding a bicycle we rely on a set of rules which are not known as such to the person following them. By a tacit integration of such factors a skilful performance is achieved.

This structure of perception and of other skills is similar to that of a scientist's skill in discovering, in "seeing" order in nature. It is one of Polanyi's main contentions that "... the capacity of scientists to perceive in nature the presence of lasting shapes differs from ordinary perception only by the fact that it can integrate shapes that ordinary perception cannot readily handle. Scientific knowing consists in discerning gestalten that indicate a true coherence in nature."[26]

There are no rules which can tell us how to make a discovery. The art of scientific discovery depends on the creative imagination and intuition of the scientist himself. He displays his skill when he sees a problem where others see none, in pursuing a solution to a problem where others see no solution. This awareness of a problem is a kind of knowledge, "a knowledge of more than we can tell". It is an intimation of something not yet fully known. Though the scientist does not immediately see the solution to his problem he has this foreknowledge of it and a growing sense of approaching a solution, which guides his selection and gathering of clues, not all of which are specifiable.[27] He does not look at these data focally but is subsidiarily aware of them as he integrates them tacitly into a conception of the reality of which they are a part. His available data enables him to intuit the pattern of a previously unknown reality. They are looked at not in themselves but as clues or pointers to the unknown and in this way we form a conception of the unknown. In this way we cross the "logical gap" between data and theory mentioned in section 2. "We may say that a scientific discovery reduces our focal awareness of observations into a subsidiary awareness of them, by shifting our attention from them to their theoretical coherence."[28] The subsidiary data are not seen in isolation in themselves, by the scientist. Using his intuition and imagination the same data are seen as part of a coherent entity.

It is for this reason that Polanyi uses the term "personal knowledge" to describe his philosophy. The person of the knower participates in all acts of understanding. Only a person can relate subsidiaries to a focus, only a person can achieve and sustain such an integration. It does not happen of its own accord. "The relation of a subsidiary to a focus is formed by the act of a person who integrates one to the other. The from-to relation lasts only so long as a person, the knower, sustains this integration."[29] The error of empiricism and positivism has been that they have tried to replace such personal participation by some explicit mechanical procedure. But this is something we cannot do. We cannot give an explicit explanation of an act of tacit integration. Firstly because many of the clues and subsidiaries are unspecifiable; secondly because we are dealing with an act of integration not deduction. Because of these indeterminacies in science there cannot therefore be any strict proof for any part of science.

There is still one further aspect of Personal Knowledge which remains to be dealt with. It is the fact that we approve our own skill by self-set standards. The skill of perception will again serve to illustrate this. There is always a measure of choice in seeing something in one way and not in another. The examples of this given by Polanyi are those of the moving ball[30] and Rubin's Vase.[31] The coherence which our tacit powers of perception achieve must satisfy us that we have truly comprehended what we see. Thus we approve our own skill. In the same way the expert biologist must acquire the skill of identifying a biological specimen. He is the acknowledged expert and he sets up standards which tell him whether or not any specimen possesses the characteristic features of a particular species. By these self-set standards he judges it to be a good or bad specimen of its kind.

The same is true of the art of discovery. The particular standards which the scientist sets himself for judging any new contribution to science we have already briefly mentioned. They are accuracy, profundity, intrinsic interest. We will look at these standards in more detail in section 5. We merely note here that the interpretation which we place on the evidence of our senses and our acceptance of that interpretation relies to a certain extent on our own convictions and approval, on standards we set ourselves. We judge it to be correct. "All understanding appreciates the intelligibility of that which it understands."[32]

All this may seem to incur the charge of subjectivity, of "authorising our own authority".[33] But this is not the case.

Polanyi answers the charge of subjectivity within a framework of commitment.[34] We have already noted how a scientist shows his originality when he sees a problem which no one else sees, and when he pursues that problem to its solution along lines of inquiry no one else suspected. To see a problem and be puzzled and intrigued by it is to believe that it must have a solution — this belief is the commitment to which Polanyi is referring. By a tacit act of integration on the part of the scientist the subsidiary clues are given their theoretical coherence. But this solution of the problem must satisfy not only the scientist himself but everyone else as well. Personal Knowledge has this universal aspect to it. A scientist believes that his theories reveal aspects of the real world, he is searching for a solution to his problem that will reveal an aspect of reality that is impersonally given and which must therefore have universal intent.

It is in this sense that the personal element of scientific knowledge is not a subjective one. A subjective feeling in us is a passive thing that happens to us. We feel fear, we feel tired. Neither of these things involves any form of commitment such as is found in scientists' belief in a solution to a problem.

Within this framework of commitment Personal Knowledge transcends the distinction between the objective and subjective. It is not a "detached" objectivity in that it involves the personal aspect which we have looked at already. It is not subjective in that it has this impersonal, universal aspect to it. Moreover this framework of commitment is absolutely necessary if we are ever to accept any piece of knowledge as true. We have come to the point where we must make up our own mind about it. An outside observer "cannot compare another person's knowledge of the truth with truth itself. He can only compare the observed person's knowledge of the truth with his own knowledge of it."[35]

The personal decision to accept the truth of a certain scientific theory is responsibly made. It is made with universal intent and therefore not arbitrarily. The scientist must conscientiously weigh up all the evidence critically and not let his speculative intuition know no bounds. "Intuitive impulses keep arising in him stimulated by some of the evidence but conflicting with other parts of it. One half of his mind keeps putting forward new claims, the other half keeps opposing them. Both these parties are blind, as either of them left to itself would lead indefinitely astray. Unfettered intuitive speculation would lead to extravagant wishful conclusions; while rigorous fulfilment of any set of critical rules would completely paralyse discovery. The conflict can be resolved only through a judicial decision by a third party standing above the contestants. The third party in the scientist's mind which transcends both his creative impulses and his critical caution, is his scientific conscience."[36] It is a personal judgement but a judgement which submits to the universal status of an external reality which we are seeking to understand.

4. Knowing and Being — The Relation of Indwelling and a Hierarchy of Levels of Reality

Polanyi's account of the structure of tacit knowing also includes the recognition of the importance of our body and of what he calls "indwelling" or "interiorization".[37] In the case of perception we are subsidiarily aware, as we have already mentioned,[38] of certain processes within our body, processes which we are not aware of in themselves. We are aware of these subsidiary processes of our body in terms of the position, shape, size and motion of an object outside of our body to which we are attending. Our body is never treated as an object in itself. We are aware of it, and of internal bodily processes, only in so far as we attend from them to something else. This gives our body a unique position in the universe. "Our body is the ultimate instrument of all our external knowledge, whether intellectual or practical. In all our working moments we are relying on our awareness of contacts of our body with things outside for attending to these things. Our body is the only thing in the world which we normally never experience as an object, but experience always in terms of the world to which we are attending from our body. It is in making this intelligent use of our bodies that we feel it to be our body and not a thing outside."[39]

In our previous discussion of tools and probes we also noted that we are subsidiarily aware of the feeling in the palm of our hands as we attend from it to what we are trying to achieve or comprehend. We treat tools and probes as an extension of our bodily equipment. We are subsidiarily aware of them, we dwell in them, we treat them as part of our own body. When we use a language we also extend our bodily equipment by dwelling in an articulate framework, by which we become more intelligent beings. Through our assimilating this articulate framework to ourselves we also assimilate our cultural heritage and dwell in it to understand experience.

All understanding, then, since all understanding is tacit knowing, is achieved by the process of indwelling, although the degree of indwelling increases as we move from the natural sciences to art and to history. In physics this means that a scientist dwells in his theory. He uses it as a "tool" of observation. Polanyi describes a scientific theory as being like a pair of spectacles. You do not examine the spectacles you look through the spectacles to examine things by them. Thus a scientist dwells in his theory and enjoys its beauty, "While he turns coldly away from disorderly, meaningless collocations of particles".[40]

As we move from physics to the biological sciences the degree of indwelling is intensified. In learning experiments performed on rats we judge their intelligent behaviour by comparing the rats' behaviour with how we ourselves would react in similar experimental conditions.

It is in the same way that we understand another person's mind. We know it by dwelling in its unspecifiable external manifestations, by dwelling in another person's outward bodily actions. This understanding of the mind means that two people mutually comprehend each other by dwelling within one another's external mental manifestations so that we have a continuous transition from the personal knowing of things to the personal encounter and exchange between minds. A mind may be less tangible than the inanimate realm of nature which physics studies but it is no less real according to Polanyi's definition of reality as being that which is able to manifest itself in the future in unexpected ways.

The learning of a skill involves a similar indwelling. A novice in trying to understand the skill of a master seeks mentally to combine his movements in the same way as the master of the skill does. A skill, like the mind itself, is a comprehensive entity which another person can know only by comprehending it through indwelling. By such indwelling the novice gets the feel of the master's skill. "We experience a man's mind as the joint meaning of his actions by dwelling in his actions from outside."[41]

Polanyi illustrates this with the skill involved in playing a game of chess.[42] We dwell in the master's moves to get the feel of his skill, we enter into his thought by repeating the games he has played. This example also serves to bring out another very important aspect of Polanyi's thought, namely that what we comprehend, for example the skilful conduct of a game of chess, has the same structure as the act of comprehension itself. This he develops into an appreciation of different levels of reality. In a game of chess we have two levels of reality. First of all there are the rules of a game of chess which govern the movements of each individual piece. Secondly you have the skilful playing of the game of chess, the skilful use of these rules, to achieve a given purpose, the winning of the game. This second level involves principles which cannot be accounted for in terms of the rules of a game of chess. In playing a game of chess we are subsidiarily aware of the rules of the game and we skilfully employ them. These two levels of reality correspond to the levels of the act of comprehension in personal knowledge, that of the comprehensive entity and its particulars. We dwell in the particulars to comprehend the given entity. So in a game of chess a novice will dwell in a master's moves to get the feel of his skill. These two levels of the act of comprehension have their foundation in the existence of distinct levels in reality itself, a stratified reality that consists of pairs of levels joined together.

The logical structure of these different levels of reality is illustrated by Polanyi by considering a machine. A machine consists of two levels, the parts of the machine which are governed by the laws of physics and chemistry, and the joint functioning of the different parts of the machine towards the achievement of a purpose as defined by the operational principles of the machine. These operational principles of a machine are not definable in terms of the laws of physics and chemistry which apply to the parts of the machine. They are rules of rightness which govern the correct achievement of the purpose of the machine. The laws of physics and chemistry which govern the parts of the machine can only account for the failure of a machine to achieve its purpose. They do however leave wide open the condition under which the parts of the machine may be made to operate together for a given purpose. These conditions are what Polanyi calls "boundary conditions". Thus the boundary conditions of the laws of mechanics and of the laws of physics and chemistry may be controlled by the operational principles which define a machine.

This two-levelled structure of a comprehensive entity like a machine corresponds to the two-levelled structure of tacit knowing. "Tacit knowing integrates the particulars of a comprehensive entity and makes us see them forming the entity. This integration recognises the higher principle at work on the boundary conditions left open by the lower principle, by mentally performing the workings of the higher principle."[43] It is in this way that we see how the machine works by understanding the meaning of its parts.

From this survey of the two levels of reality of a machine Polanyi goes on to apply the principle of boundary conditions and a hierarchy of different levels of reality to living beings. The first of the hierarchy of levels in living beings is the level of being itself, the level of taxonomy and morphology. In our study of the typical shapes of living beings we classify them according to standards which we ourselves believe to be true. This classification also involves our personal appraisal of an achievement, that is, whether a specimen is normal. The second level of biological achievement passes from the study of typical shapes, the level of being, to the level of growing and of coming into being. This level embraces morphogenesis and embryology. Again there is involved at this level our personal appraisal of the success or failure of these processes. The third level is the level of physiology, the study of the proper functioning of the organism and of the vegetative functions of the various organs of the body. It seeks to discover the operational principles which will account for the healthy functioning of the organism. This involves a personal appraisal of health and disease. The next level is that of the active perceiving individual. In the conscious action of feeding for example we have an action which we can judge to be correct or mistaken, to be right or wrong feeding. This necessitates our recognition at this level of an appetitive/perceptive centre in the individual which is capable of correct or mistaken decisions. This is clearly seen when we come to study the intelligence of rats in learning experiments which we set to them. We try to imagine what we would do in the rats' position and deem the rats' actions as intelligent or not accordingly. Perception also requires a decision to be taken as to whether or not what we see is mistaken or not. We have here in contrast to the two-storied logical structure of inanimate science a three-storied structure. The individual under observation is itself making decisions, which we can judge to be mistaken or correct. A science dealing with living beings therefore is logically different from a science dealing with inanimate things.

In these ascending levels of biological achievement we notice two things. First of all there is a greater degree of indwelling. Initially we are concerned with the appraisal of an achievement, when we come to the study of the intelligence of an animal our appreciation of its intelligence is "convivial", there is a link between its person and ours. This is something which we have already noted. Secondly we can see the principle of marginal control at work. "The vegetative system which sustains life at rest leaves open the possibility of bodily movement by means of muscular action. This level leaves open the possibility of integration into innate patterns of behaviour. This leaves open the shaping by intelligence . . .".[44] Just as a machine cannot be explained in terms of the principles of the lower level of physics and chemistry so too with the different levels of reality in biology. The principles governing a lower level can be shaped by the principle of a higher level.

This study of a hierarchy of levels in biology leads on to a very important aspect of Polanyi's thought, an aspect which will be seen to be especially relevant for our study of authority and tradition which we will look at in our next section. When we move on from the level of an active perceiving individual we come to man and to man's knowledge. We have noted above that when observing the inanimate level of existence we have two "logical" levels, one for the observer one for the object. A third level comes into existence when we observe an animal which is itself an observer. The relationship between observer and object becomes increasingly more convivial until with the study of man we encounter, not observe, another's mind. The understanding of life at all levels has involved an appreciation of what Polanyi calls biological achievements by standards we set to the organism by ourselves. But when we come to study another person and their knowledge by standards we set to them we find ourselves in a situation where we are reflecting on our own knowledge. We have "a confluence of an extended biology with the theory of knowledge".[45] This brings us back to the point we have previously discussed that we must acknowledge our powers of thought for knowing the truth, knowledge which we judge to be true by standards set to it by ourselves.

In this survey of the ascending levels of biology we have reached the position of equality between ourselves and the person whose knowledge we are examining. Here we are in the position of judging the truth of another person's knowledge but also in the position of having to criticise and examine our own understanding. We encounter, not observe, another person's mind, and we apply to him the same standards we accept ourselves. Similarly, the person whose mind we encounter also encounters our mind and applies to us his standards. For any meaningful dialogue to take place these standards must be the same. We begin to see now in this context the need in science for an authoritative traditional framework. We need to judge each other's knowledge in the light of the same standards, in the light of the standards of the scientific community of which they are a part. "Take two scientists discussing a problem of science on an equal footing. Each will rely on standards which he believes to be obligatory both for himself and the other. They both rely on a whole system of facts and values accepted by science. They both trust each other to accept these values and there is a bond of mutual trust between them which is but one link in the vast network of confidence between thousands of scientists in different specialities which may be said to accept certain values and facts as scientifically valid."[46] We will look more closely at this consensus which forms the authority of science and how it operates a little further on, but for the moment we see it as an extension of the biological achievements to which Polanyi's survey of the ascending levels of living being has led us. As we mentioned before man comes into existence mentally as he dwells in the articulate framework of his culture. This mental life was what was called the "noosphere". Here man seeks the satisfaction of his intellectual passions, to satisfy his desire for truth and for things of intrinsic worth. Unlike our bodily passions man's mental passions are not self-satisfying. The truth he finds he hopes will satisfy others and will be universally valid. Man's power of thought we believe reveals aspects of reality and has the ability to reveal and know the truth and so enrich other minds. So with science. Here we have a knowledge of our world accepted as valid by the mutual consensus of many scientists, cultivated by them and accepted by the public on their authority as being true. The articulate framework of the scientific community is the articulate record of the works and discoveries of the past masters of science. Their knowledge is embodied in this articulate framework. The whole life of the scientific community is based on the assumption that the standards passed on to us by our masters are right and that the knowledge embodied in a scientific tradition descending from them is valid, and more than that since it reveals aspects of reality the facts and values of science bear on an as yet unrevealed reality. We will also enlarge on this a little further on but we note here that this authoritative traditional framework of science depends on the belief in the ability of man's thought to discover, disclose and uncover aspects of reality, to recognise and appreciate truth when confronted by it.

This leads on to the last level in Polanyi's survey of Biology. In accepting an articulate tradition we pass on from an encounter of equals on an equal footing to our recognition of great men in the scientific tradition, men whose teaching and standards we follow. All articulate records are the records of the work of the great men of science and we trust ourselves to these men and follow them. "Our adherence to the common beliefs and standards on which intellectual exchanges within a culture depend, appears then equivalent to our adherence to the same masters as fountains of authority."[47] We look up to these past masters and we dedicate ourselves to the standards and knowledge they have formulated. We let them guide us. "Our possession of knowledge is seen to consist in an act of understanding and submission."[48] Thus the dialogue between two scientists on an equal footing, involves their submission to the standards and authority of the tradition established by the famous men of the scientific community.

This whole process of acquiring scientific knowledge requires our submission to the works of past masters whose thoughts and works and deeds we study. Yet we are still free to choose which masters we follow. This is especially seen for example in times of scientific controversy. The decision is ours in accepting them as our masters and the tradition of science as authoritative. No authority can teach us whom to follow. For example in the various rivals which there are to the scientific understanding of the world, astrology and the magical, no one can teach us which is right. We ourselves choose to reject the rival interpretations as being intellectually less satisfying, and choose to follow the scientific interpretation of events, and the tradition as laid down by great scientists. We apprentice ourselves to them and trust and accept their authority. "By applying his thoughts or deed as our standards for judging the rightness of our own thoughts and deeds, we surrender our person for the sake of becoming more satisfying to ourselves in the light of these standards. This act is irreversible and also a-critical, since we cannot judge the rightness of our standards in the sense in which we judge other things in the light of these standards."[49]

The validity of science then, over against other interpretations of nature, rests upon a preference of interpretation as developed over the centuries by our civilisation, by a scientific community bound together by loyalty to a common tradition. This is a preference, however, which is not subjective but is based on the belief in the power of human thought to create theories which reveal aspects of reality and judge their truth and intrinsic worth.

5. *The Nature and Role of Authority and Tradition in Science*

In the first section of this introduction we noted how a false ideal of science had become unconditionally accepted. This false ideal identified truth with that which could be impersonally demonstrated, leaving no justification for any acceptance of knowledge on the grounds of authority or tradition. The exercise of authority, on the basis of this false ideal, will tend to appear as being bigoted. Traditional values and the framework of knowledge transmitted from generation to generation will appear to be arbitrary. We cannot hold values or unproven beliefs, only real knowledge "objectively" demonstrated. Such is the view put forward by Bertrand Russell: "The triumphs of science are due to the substitution of observation and inference for authority. Every attempt to revive authority in intellectual matters is a retrograde step. And it is part of the scientific attitude that the pronouncements of science do not claim to be certain, but only the most probable on the basis of present evidence. One of the great benefits that science confers upon those who understand its spirit is that it enables them to live without the delusive support of subjective authority."[50]

The attitude of doubt and scepticism towards tradition and authority and belief, has not only affected the realm of science but has spilled over into society and caused a civic predicament.[51] This civic predicament was indeed one of the things which prompted Polanyi to carry out his reform of our conception of knowledge and truth.[52] While it is true, that in its formative period, science was based on a rejection of authority, this must be seen in the context of that period. It was a rejection of all external authority which at that time threatened the pursuit of science. "Once these opponents were defeated, however, the slogan remained and came to imply that the pursuit of science required the repudiation of *all* authority and *all* tradition. It became very misleading at this point, since, as we shall see, the pursuit of science certainly does not and cannot repudiate all authority and tradition."[53] Indeed a close study of the scientific process will show that science itself is based on authority and tradition, and that the particular view of science, which sees science as rejecting all authority and tradition is completely unfounded.

We have seen that the art of scientific discovery is a skill similar to that of perception. We have seen too how we cannot make completely explicit our scientific knowledge. As Polanyi points out, "We know more than we can tell". There are no explicit rules that can be laid down and passed on to others that will tell us how to make a discovery, how to recognise a problem and which ones are worth pursuing, what evidence should or should not be accepted for a particular theory. There is much that cannot be made explicit because it lies at the level of feelings about the fitness of things, the beauty of a theory, which we noted in section 2. This is the "personal" element in our knowledge which we have discussed above.

Like other skills the skill of scientific discovery can only be passed on by example from one generation to the next and learned to a certain extent by imitation. We cannot explicitly lay down the premisses of science as if they could be formally passed on to the next generation. Scientific discovery shares with all other skills the fact that the premisses of the skill are not known focally, or properly understood by us, prior to the performance of that skill. We must learn the skill first. We can for example, as we have already mentioned, ride a bike without knowing how we do it, without knowing the "premisses" focally that underlie that skill, we know them subsidiarily as part of the mastering of that skill and all explicit knowledge, all maxims or rules of art cannot be known prior to the performance of the skill. "Rules of art can be useful, but they do not determine the practice of an art; they are maxims, which can serve as a guide to an art only if they can be integrated into the practical knowledge of the art. They cannot replace this knowledge."[54]

So then there are no rules or precepts that can be handed on to us that will tell us how to undertake scientific research and choose our problems and reject or accept the various clues that arise in the course of our investigation. We can learn a skill, and receive the premisses of science only in imitating someone who is a recognised authority in science, and by practice. "Since an art cannot be precisely defined, it can be transmitted only by examples of the practice which embodies it. He who would learn from a master by watching him must trust his example. He must recognise as authoritative the art which he wishes to learn and those of whom he would learn it."[55]

It is in this way that the premisses of science are passed on, by learning from the example of a recognised authority in the art of scientific discovery, and so to accept an "artistic" tradition in which are embedded these premisses. It is for this reason that very often little scientific progress is made in non-European or American countries where there is no scientific tradition, in which there is no authoritative figure from whom to learn the skills of scientific research.[56] It is for this reason too that very often "great scientists follow great masters as apprentices".[57]

Polanyi uses the example of learning a language to illustrate further the way in which the premisses of thought are transmitted from one generation to the next.[58] A child learns to speak by imitating adults. Every word that is used must be remembered in the various contexts in which it is used. The child has to learn what we mean by words, and attend from the words we use to that to which they refer. This is a tacit skill learned by imitation and by practice and involves in the first instance the acceptance and belief by the child that the words the adult is using are meaningful. There are no explicit rules by which a language is learned.

So too in science. We must accept first of all that science

is a meaningful and valid system of thought. We must believe in science; we must believe in order to know. And since there are no rules to guide us in scientific discovery we must learn these skills from the example of the skill of a master. As we learn these skills we receive subsidiarily the premisses of science, which can then later, by analysis of the skill, be known focally.

"We have seen that tacit knowledge dwells in our awareness of particulars while bearing on an entity which the particulars jointly constitute. In order to share this indwelling the pupil has to accept the fact that a teaching which appears meaningless to start with has in fact a meaning which can be discovered by hitting on the same kind of indwelling as the teacher is practising. Such an effort is based on accepting the teacher's authority."[59]

Let us enlarge now on these two aspects, namely that we must accept the authority of those from whom we would learn, and trust them; and also we must accept science as a valid system of thought, we must believe in science.

Polanyi points out that in the process of learning science the pupil goes through three stages.[60] There is what he learns at school then at university and then as an independent scientist. But it is only at this final stage that the scientist becomes fully introduced to the premisses of science. He learns by watching the way a master undertakes his scientific investigations. "In the great schools of research are fostered the most vital premisses of scientific discovery. A master's daily labours will reveal these to the intelligent student and impart to him also some of the master's personal intuitions by which his work is guided. The way he chooses problems, selects a technique, reacts to new clues and to unforeseen difficulties, discusses other scientists' work, and keeps speculating all the time about a hundred possibilities which are never to materialise, may transmit a reflection at least of his essential visions."[61]

But it is only as he becomes an independent scientist actively engaged in research and discovery that he comes nearest to the premisses of science. Even if the scientific achievements of this century were analysed we would only know the past premisses of science. The premisses of science at this actual moment in time are as Polanyi points out the ones being formed in the mind of the scientist actively engaged in research and which guide his research.[62]

It is by learning from such a person, by following his example, that we learn of the premisses of science. So we must accept and trust the authority of such a person and the whole process of learning relies on just such an acceptance by us of his authority. But this authority, this teaching and practice, to which we submit as we learn, is there to bring us into contact with the reality of nature. Indeed the whole process of learning and of assimilating the premisses of science is very similar to the actual structure of discovery itself when we originally intuited a hidden pattern in nature. The doctrine and the theory and the experiments we perform, which we find in textbooks and which we accept on authority, are there as clues which we tacitly integrate and so intuit the rationality in nature to which they point us.[63]

Eventually the student will begin to rely less on the authority of his teacher as he advances from being a student to an independent scientist. He will begin to rely more on his own judgement and intuition and conscience. The tradition and authority of science have achieved their purpose of bringing him into contact with the reality of nature and he seeks to advance such knowledge and make new discoveries through his own intuitive contact with this reality.

So then we can see that the domain of science itself depends on our acceptance of authority which may on occasions have to be enforced by discipline.[64]

We come now to the second aspect, that we must believe in science as a valid system of thought, that it is meaningful. We believe in science and in so doing we reject belief in any other system of thought such as astrology. Not all people share this belief in science. Some even today still believe in astrology and find evidence to support their beliefs, and so such beliefs remain to challenge the authority of science.

Why this should be so is illustrated by Polanyi's reference to the Azande,[65] which illustrates the circularity or stability of a conceptual system. We believe in a naturalistic system of thought rather than a magical or superstitious one. But the logical structure of both is the same. Both operate by the principle of "suppressed nucleation" and have an "epicyclical structure". "Our attribution of truth to any particular stable alternative is a fiduciary act which cannot be analysed in non-committal terms."[66] We do not believe such superstitious beliefs as the Azande. We disregard them and replace them by naturalistic ones. We do so because we believe science to be true — to bring us into contact with the reality of nature.

Yet even within science itself we are faced with a similar situation during times of scientific controversy when several stable conceptual systems rival each other. For example the controversy over fermentation[67] or the controversy between the Ptolemaic system and the Copernican. Each of these stable systems accounts for certain facts and also accounts for why other facts are neglected or explained away. Each has its own language and conceptual framework and each accounts for the facts equally well. There is a logical gap between both controversial conceptual systems and no formal argument within one framework of thought can convince another person working in another conceptual framework. This has to be done by persuasion, by a "conversion" from one to the other.

But this raises the question as to who decides the outcome of such controversies. Who decides what "science" is science? We have to believe in science first of all if we are eventually to understand it, what science do we believe and on whose authority?

Let us look first of all at the fact that though each scientist displays his own originality and uses his own personal judgement yet nonetheless there is in the scientific community a harmony, and agreement amongst scientists as to what is "science". This harmony is seen in the fact that each scientist depends on the results of other scientists. The scientific community is a closely knit and coordinated community. This coordination is achieved by what Polanyi calls "... the adjustment of the efforts of each to the hitherto achieved results of the others. We may call this a coordination by mutual adjustment of independent initiatives — of initiatives which are coordinated because each takes into account all the other initiatives operating within the same system."[68] The work of the scientific community is not guided by a central authority but by the self-coordination of individual scientists to each others results.

Polanyi illustrates this principle with several examples. He compares the coordination achieved between scientists with the coordination achieved in a market place between producer and consumer to make supply meet demand.[69] This coordination is achieved through the use of money and the price of goods, which are adjusted in terms of scarcity or glut. A similar coordination is seen in the solving of a jigsaw puzzle.[70] Here the progress made by a group of helpers will be greater than the progress made by an individual when the helpers are allowed to act independently on their own initiatives while at the same time they watch what the others are doing and adjust their individual actions appropriately. No central authority directing the team could achieve a quicker solution. Such an intervention would bring no real improvement on the progress made by one individual.

This is how Polanyi sees the organisation of the scientific community. Each scientist acts on his own initiative and shows his own originality but nonetheless he takes into account all the achievements and results of other scientists. If he worked in isolation no scientific progress would be made. If he were to be guided by some central authority, or if science were to be planned this would limit the scientist's own originality and intuition in recognising and pursuing the solutions to problems and so severely handicap the progress of science. He coordinates what he himself is doing to what other scientists have done.

This is the point where a scientist is guided by professional standards and by his commitment to a "transcendant spiritual reality" in knowing which contributions and results of other scientists are to be accepted as additions to our knowledge and which are not. These standards also help him to choose a good problem and to recognise the importance of any solution he hits upon. It is a scientist's allegiance to these ideals and standards that keeps out all that is trivial and of little scientific interest or value from what is accepted as science. This is illustrated on a number of occasions by Polanyi in the appointment of scientists to academic posts and in the way contributions to scientific journals are assessed.[71]

Each contribution is assessed by three standards before becoming accepted, that of Plausibility, Scientific Value and Originality.

The first standard is that of plausibility which Polanyi illustrates with several examples; for example that of Lord Rayleigh,[72] the Velikovsky Affair,[73] the gestation period in animals.[74] Here certain contributions to science are seen as not being plausible in the light of the existing and accepted conceptual framework of science. Our conception of the nature of things suggests to us that such contributions to science are absurd and here authority prevails against the "facts" on which these contributions are based. This assessment of plausibility relies on a scientist's intuition and is therefore a tacit judgement on the part of the scientist. This also means that such tacit judgements, since they rely on the personal judgement of the scientist, may not always be correct, as is illustrated by an example out of Polanyi's own scientific career.[75]

The second standard by which contributions to science are judged is that of Scientific Value which is divided into three components, accuracy, systematic importance and intrinsic interest.[76] These two standards produce an inevitable conflict within science. The standards of scientific value and of plausibility tend to make all knowledge conform to the existing conception of the nature of things, whereas the third standard of originality may in fact mean a rejection or a change in the existing framework of thought. This is illustrated for example in the Copernican Revolution. Scientific authority tends to enforce conformity to the existing framework of thought yet at the same time allows room for original contributions to be acknowledged.

This two-fold function of scientific standards of enforcing the established teaching of science yet fostering originality derives from the scientist's belief that his theories reveal an aspect of reality. We have already noted how Polanyi sees the structure of scientific discovery as being similar to that of perception, whereby the existing knowledge in science serves as clues to future discoveries and supplies new insights into the reality of nature of which we have grasped a part. New discoveries and new problems are suggested by existing knowledge. "We know more than we can tell." This is seen by the fact that "Copernicus and Kepler told Newton where to find discoveries unthinkable to themselves".[77] This belief in scientific theories revealing aspects of reality, explains why the application of the above standards produces conformity to the existing framework of knowledge yet still allows the appearance of originality, and why science retains its harmony and unity down through the years.

Yet we still have not fully answered the question as to who decides what "science" is science, who applies the above standards. Here we must answer that the scientific community as a whole does this through what Polanyi calls "scientific opinion". No one scientist can possibly know more than only a tiny fraction of all scientific knowledge. He is competent to judge contributions in his own field of study and also in closely related fields. So no one scientist can decide on this matter, neither can any central authority. But because of the fact that each scientist is competent to judge contributions in the fields of study overlapping with his own and because scientific standards are the same throughout all the branches of science, there is a "principle of overlapping neighbourhoods",[78] where each individual scientist judges and criticises contributions in a neighbouring field. The authority of science is not therefore exercised by one man or by a central authority but is an authority established between scientists not over and above them, by a world wide community of verifiers.

The authority of this scientific opinion which is established between scientists determines what is and what is not science, and it is to this scientific opinion that the student has to submit in the process of learning.

It has to be remembered too that the function of the above authority goes far beyond any mere confirmation of the facts asserted by science. There is as Polanyi points out no such thing as mere facts.[79] This is clearly seen in times of scientific controversy when each side in the conflict accepts certain facts which the other ignores or explains away. Science is not a mere collection of facts based on their scientific interpretation. All genuine facts are theory laden. It is this system of thought that is accepted and endorsed by scientific opinion. A scientific fact is one that has been accepted as such by scientific opinion and its acceptance is based on evidence for it and because it appears plausible in the light of the existing conceptual framework of science. In accepting the authority of science we accept these value judgements.

In the light of what we have looked at so far we can now begin to understand what Polanyi means when he says "The authority of science is essentially traditional".[80] The student in learning by the example of a master is accepting a tradition which has been passed on down through the years and which has developed through the years, and he himself becomes in turn a representative of that tradition. Both his appreciation of scientific merit and understanding of the method of scientific enquiry are based on a tradition which each generation accepts and develops. But this tradition is a "dynamic" tradition, it cultivates originality while at the same time exercising its powers of authority over the student. It is a tradition that undergoes a continual process of self-renewal. This is the outcome of the scientist's belief that his theories reveal aspects of reality which may reveal itself in new and surprising ways in future discoveries. There is an objective "spiritual reality" embodied in this tradition and transcending it.

While the student must accept the authority of such a tradition nonetheless as he intuits the rationality in nature through his schooling in this tradition he himself has to decide, in the light of his intuition of the rationality in nature, what parts of that tradition need to be changed. He himself in the light of present scientific controversies has perhaps to rethink the lessons and the outcome of past scientific controversies.[81] But even when doing this he is taking his stand within that tradition and trying to convince other scientists by appealing from tradition as it is to the tradition as it ought to be.[82]

"Since a dynamic orthodoxy claims to be a guide in search of truth, it implicitly grants the right to opposition in the name of truth."[83]

So it is that the premisses of science are embodied in a tradition that is passed on by example from master to pupil. Each scientist must be loyal and be dedicated to the scientific standards and ideals we have mentioned and be prepared to oppose scientific opinion where necessary in upholding those ideals. It is only in this way that science can continue to exist, only in so far as each scientist accepts and is informed by one tradition and is wholly committed to these scientific ideals and believes in science as revealing aspects of reality.

"When the premisses of science are held in common by the scientific community each must subscribe to them by an act of devotion. These premisses form not merely a guide to intuition but also a guide to conscience; they are not merely indicative, but also normative. The tradition of science, it would seem, must be upheld as an unconditional demand if it is to be upheld at all. It can be made use of by scientists only if they place themselves at its service. It is a spiritual reality which stands over them and compels their allegiance."[84]

Polanyi's heuristic philosophy has several contributions to make to theology. Theology itself has been traditionally described as one of the "sciences", whose task it is to apprehend and understand and speak of God. But theology itself like many other areas of knowledge has become involved in and affected by the objective ideal of knowledge which has fostered the schism between knowing and being, mind and body, reason and experience, facts and values, the knower and the known, the subjective and the objective. In this sense Polanyi's philosophy is an uncomfortable ally for theology as Richard Gelwick points out.[85] It challenges such an objective ideal of knowledge wherever it is pressed into the service of theology.

In our previous discussion we have seen how in science itself we must believe in order to understand. Science requires of us that we believe knowledge can be ours, that we trust ourselves to the authority and example of a master who is a representative of the scientific tradition. Such traditionalism "is based on a deeper insight into the nature of knowledge and of the communication of knowledge than is a scientific rationalism that would permit us only to believe explicit statements based on tangible data and derived from these by a formal inference, open to repeated testing".[86]

This affirmation of an authoritative tradition in science, based on the rejection of the objectivist ideal of knowledge and its replacement by personal knowledge, has a bearing on our understanding of the role and nature of tradition and authority in theology. Polanyi himself recognises the bearing which the reaffirmation of traditionalism has on religious thought.[87] He sees the enfeebled authority of revealed religion being revived by the overthrow of the sceptical attitude inherent in objectivism, which his philosophy achieves. The prevailing sceptical attitude in our culture has arisen out of a false ideal of science and can only be overcome by a re-examination of the nature of science. It is just such a reappraisal of science that Polanyi carries through so helping to re-establish our confidence in the role of authority and tradition in theology.

Perhaps we might now more specifically deal with several areas where Michael Polanyi's understanding of authority and tradition is of relevance to our understanding of authority and tradition in Theology.

CHAPTER II

THE AUTHORITY OF THE BIBLE

MICHAEL POLANYI summarises his fiduciary programme in this way, "I believe that in spite of the hazards involved I am called upon to search for the truth and state my findings".[88] His philosophy is one that involves "the declaration of my ultimate beliefs".[89] He declares that "We can voice our ultimate convictions only from within our convictions — from within the whole system of acceptances that are logically prior to any particular piece of knowledge. If an ultimate logical level is to be attained and made explicit this must be a declaration of my personal beliefs. I believe that the purpose of philosophic reflection consists in bringing to light, and affirming as my own, the beliefs implied in such of my thoughts and practices as I believe to be valid; that I must aim at discovering what I truly believe in and at formulating the conviction which I find myself holding."[90] All of this derives from Polanyi's understanding of Personal Knowledge as meaning that commitment which arises as a personal decision accepting something that is believed to be impersonally given. He is putting forward the claim that the knowing person can recognise an objective reality, can discover and know the truth. Indeed it is this assumption, namely the existence of truth, the existence of an objective reality, that drives forward the scientific discoverer, to search for that truth and state his findings. In Polanyi's fiduciary programme there are no grounds of justification whatsoever for any attempt to define truth in impersonal terms. As we noted in the introduction to Polanyi's thought we cannot compare a person's knowledge of the truth with truth itself, as if truth was something that was impersonally given. We can only compare another person's knowledge of the truth with our own knowledge of it. "Truth is something that can be thought of only by believing it."[91] The act by which we search for the truth and declare our findings is an act of an inherently personal character. Truth is not something that is impersonally given, it is something that only a person is capable of apprehending.

This search for truth and the stating of our findings involves a procedure that is necessarily a circular one. In accepting the scientific explanation of phenomena we enter into the proper circularity of any coherent framework of thought which contains one or a number of ultimate beliefs. For example, the scientist believes that there is an intelligible order in the world, objectively given. The conceptions we form of that order are given due authority over us because the rationality of our conceptions is believed to be due to their being in genuine contact with reality. But there is no means by which we can prove that there is intelligible order in the world objectively given. There is no experiment we can perform to demonstrate it. Yet it must be presupposed before there can be any science at all. The scientist believes in an ordered universe and seeks to make that order explicit. Such a procedure must assume that which it is trying to prove. We must believe in that order, in the ability of our minds to grasp it and in the fact that the conceptions we form of it intimate in their own rationality the prior rationality of this ordered world we live in. We must believe in order to understand. These ultimate convictions or ultimate beliefs cannot be justified on any grounds other than that they have an authority of their own, a persuasive appeal, that convinces us of their truth.

Our acceptance of science, therefore, involves an act of personal commitment to certain ultimate beliefs which we are persuaded to accept under the impact which the intrinsic intelligibility of objective reality itself makes upon us, as we are confronted by it. The scientific interpretation of nature wins our acceptance, because of its own power to convince and persuade us of its truth. We have to abandon all efforts to find strict criteria of truth and strict procedures for arriving at the truth, and accept commitment as the only way in which we can believe something to be true.[92] We commit ourselves by our own personal judgement, responsibility made, to certain beliefs. In the case of the scientific interpretation of phenomena our decision to commit ourselves to the scientific tradition, being a commitment to certain ultimate beliefs, means that as such there is no objective demonstration by which we can judge the correctness or otherwise of our decision. We believe that there is an intelligible order in the universe, that science is formed under the impact which the actual rational structures of the real world make upon our minds as we seek to understand it. Belief here then comes to mean that on our part it is "the subjective pole of commitment to objective reality, but intelligent commitment to an objectively intelligible reality".[93] Belief means assent given by a knowing believing person as he turns away from himself to the reality which confronts him and which convinces us out of itself and on no other grounds, of its own inherent intelligibility. We are confronted by the reality of the ordered world around us which we must either shut our eyes to (which would be irrational since we would not be behaving in accordance with what is there) or accept and acknowledge as being independent of us and of our knowing of it, a reality that calls forth our responsible commitment (which would be rational behaviour, behaving in accordance with what is there).

As Polanyi makes clear the circularity of the scientific interpretation of the world must be consistent with itself. When, in times of scientific controversy, a new paradigm has emerged, this has often involved conversion from one self-consistent framework of thought to another, or of a reconstruction of our framework to make it more consistent.

In times of scientific controversy when faced by two self-consistent frameworks of thought the conflict between them can only be resolved by our own decision, responsibly made, to accept one as being intellectually more satisfying. We are converted from one framework of thought to another because we recognise that it reflects the intelligibility of objective reality in a more satisfying way, and as such is true. Again it is the case of searching for the truth and stating our findings, of giving voice to our convictions.

The authority of science therefore resides in itself or rather in its correlation with reality. The authority of the scientific interpretation of the world rests on its own intrinsic powers to convince us out of itself of its truth. There is no higher authority to which we can go to judge or prove the scientific interpretation of the world and its ultimate beliefs. In the last resort it is reality itself, to which science is correlated, that must be the judge of what is true or false. Its authority rests on its own power to convice us out of itself, on the impact which the inherent intelligibility of the universe makes on our minds as we seek to understand it.

Polanyi's analysis of the circularity of the scientific framework of thought in terms of "commitment" and "personal knowledge" prepares the way for the acceptance of the authority of a revealed religion.[94] Indeed Polanyi's understanding of truth as that which can be thought of only by believing in it is parallel to the thought of the Fathers' on the concept of the truth.[95] The patristic concept of truth was that truth was "that which is what it is and discloses itself to be",[96] and which "shines in the light of its own evidence and forces upon us an assent and a consent in relation to it".[97] In other words belief is the assent from us to the self-evidencing reality of things and there is only this one basic way of knowing whether we are referring to theology or to the natural sciences. Account was taken of the fact that though there was only one basic way of knowing, in which we have to think in the way we are compelled to think by the nature of the reality we are studying, yet in different fields of reality we operate with modes of investigation and verification appropriate to the nature of that reality. The nature of the subject-matter determines how knowledge of it is to be obtained. When this concept of truth is applied to our knowledge of God we must say that "the truth of God is that He is who He is and that He reveals who He is as He is".[98] We are confronted with the self-evidencing Truth of God to which we can only give our assent, our response of faith and love, the only rational response.

We are faced in theology with the same kind of proper circularity which Polanyi has shown to exist in science and in any coherent framework of thought containing certain ultimate beliefs. For the theologian certain historical acts in which God has revealed Himself are our ultimate beliefs. We must first of all state our convictions. We can only say that in Jesus Christ we are confronted with the reality of God's self-revelation. The ultimate reality of God's self-revelation in the incarnate life of Jesus Christ is something that must authoritatively speak for itself. There is no way in which we can objectively (i.e. in a detached way) demonstrate the Church's belief in the self-revelation of God in Jesus Christ. It must convince us out of itself. Furthermore we must try to understand God's self-revelation in Jesus Christ within a framework of thought which presupposes that revelation and its reality. Such an admission of circularity is justified as Polanyi points out, "only by my conviction that in so far as I express my utmost understanding of my intellectual responsibilities as my own personsl belief, I may rest assured of having fulfilled the ultimate requirements of self-criticism; that indeed I am obliged to form such personal beliefs and can hold them in a responsible manner, even though I recognise that such a claim can have no other justification than such as it derives from being declared in the very terms which it endorses".[99] So it is that just as in science, where a scientist seeks to make explicit the order inherent in the world within a framework of thought that presupposes that order, so too must the theologian seek, for example, to understand the Resurrection of Christ in a framework of thought that presupposes the Resurrection. "As such, then, the incarnation and the resurrection together form the basic framework in the interaction of God and mankind in space and time, within which the whole gospel is to be interpreted and understood. But they are *ultimates,* carrying their own authority and calling for the intelligent commitment of belief, and providing the irreducible ground upon which continuing rational enquiry and theological formulation take place."[100] It is a case of believing in order to understand. But as we have also seen this is true of science. We must believe in science too, to understand it. It is a case of submitting ourselves to the authority of the truth itself, in its own self-evidencing intelligibility which confronts us and calls forth our acceptance or rejection.

The Resurrection and Incarnation of Jesus Christ are only two of a number of ultimate beliefs in the Christian faith which must be allowed to convince us out of themselves, bearing their own proof of the reality of God's self-revelation in Jesus Christ. In theology as in science we are dealing with self-attesting, self-authenticating truth, the main difference being that in theology we have to do with the Being of God whereas in science we have to do with created being and also that the ultimate beliefs in theology are certain historical events in which God has Revealed Himself and in which He is experienced. While Polanyi does not state that such ultimate beliefs may arise out of history, Thomas Langford suggests that nonetheless his methodology is open to this suggestion.[101]

We are then faced in theology with the Ultimate authority of God in Jesus Christ, an authority which we cannot control or manipulate, or judge by any higher authority or test by any procedure for arriving at its truth. In theology as in science we are confronted by the authority of self-attesting self-authenticating truth which calls forth from us our responsible commitment.

Yet we must now go on to add that we come face to face with this ultimate authority of God's self-revelation only in and through the Biblical witness of the Prophets and Apostles. The Biblical witness of the Apostles for example has arisen in response to God's self-revelation in Christ, it has taken its distinctive form under the creative impact of

Christ Himself, and is meant to point us beyond itself to Him, to that same Reality. We are faced in the Biblical tradition with an articulate framework of thought brought into being by God's self-revelation in Christ, by the fact and reality of that revelation. The intention of the Prophets and Apostles was to speak of it, of what they had heard and seen, that we through their witness might be directed to hear that same Word of God. As Karl Barth points out, their witness is the primary source of our understanding of God, as they and only they, the Apostles and Prophets, were present and in immediate relationship to His revelation, whereas we are not. It is only through their witness that we come to hear and believe.

The question of the authority of the Bible vis-à-vis the ultimate authority of God's own self-revelation in Christ to which it points us raises several questions on which Polanyi's philosophy sheds some light. The above discussion has involved the tacit assumption, namely that we know that the Biblical witness, the Biblical canon, is authoritative and to be accepted and respected. But how do we know this? In answer to this question we must again say that we enter the logical circle of self-asserting, self-attesting truth. The Bible is a self-authenticating witness to divine revelation for in and through its witness we can only say that we are indeed brought before the ultimate authority of God. "The Bible must be known as the Word of *God* if it is to be *known* as the Word of God. The doctrine of Holy Scripture in the Evangelical Church is that this logical circle is the circle of self-asserting, self-attesting truth into which it is equally impossible to enter as it is to emerge from it: the circle of our freedom which as such is also the circle of our captivity. . . . When the Evangelical Churches of the Reformation and later were asked by their Roman adversaries how the divine authority of Scripture could be known and believed by men without being guaranteed by the authority of the Church, the Evangelical theologians gave the hard but only possible answer that the authority of Holy Scripture was grounded only in itself and not in the judgement of men."[102]

In the Bible we come face to face with the Truth of the Living God which we can only know and recognise as such. The authority of the Bible in the Church is not something that is established by the Church by some criteria for recognising its truth, for the Church has no prior knowledge of revelation by which it can judge the Biblical Revelation. It does not give authority to the Bible but only establishes and recognises its authority. The canon of Holy Scripture is not formed by the Church, it is confirmed by it. "Holy Scripture is the Word of God and makes itself known as such."[103] We can only search for the Truth of God in the Biblical witness and state our findings and give voice to our convictions that there indeed we are confronted by the Truth of God. That is why according to Karl Barth it has authority in and over the Church. The authority of the Bible is not an authority to be placed alongside the authority of the Church. "The Church cannot evade Scripture. It cannot try to appeal past it directly to God, to Christ or to the Holy Spirit. It cannot assess and adjudge Scripture from a view of revelation gained apart from Scripture and not related to it. . . . It cannot establish from any possession of revelation the fact and the extent that Scripture too is a source of revelation."[104] We receive in obedience the authoritative witness of the Apostles. Only the Apostles stand in immediate relationship to God's self-revelation. It is only through their witness that the Church is brought face to face with the same ultimate authority of God and so therefore absolute authority must be given to Holy Scripture in the Church and over the Church. The Church has no direct access to God's self-revelation as the Apostles had, it can only accept and reproduce their witness. Yet as Karl Barth points out, "It is not the book and letter, but the voice of the men apprehended through the book and the letter, and in the voice of these men the voice of him who called them to speak, which is authority in the Church".[105] We can distinguish between seeing Jesus and hearing his Apostles and yet we cannot separate the two and try to have one without the other. We must accept the Biblical witness as authoritative for us since it is through its witness that we are confronted with the ultimate authority of God.

Polanyi's account of indwelling is perhaps of relevance to our consideration of this particular point. The Biblical witness has arisen in response to God's self-revelation in Jesus Christ. We have therefore to dwell in the Biblical tradition, in the words of the Apostles, seeking to achieve by an integrative act the same heuristic vision of God as it was the Apostles' intention to convey. We hear their words but we don't focus our attention on them; we indwell them as subsidiary clues and attend from them to the Reality of God's self-revelation in Jesus Christ, which first gave rise to them and gave to them their distinctive form and shape. How we gain such an understanding of God through in integrative indwelling of the Apostolic witness we cannot tell. But that is also true of all other areas of knowledge as we have already mentioned. We seek to follow the witness of the Apostles till we encounter the Reality that gave rise to that witness, we attend from their words to that reality. Their witness points us to God's self-revelation and in the Spirit we are able to speak of it. If we follow their witness and find nothing at the place to which their witness points there can be no understanding of the Bible, such understanding comes only as we are confronted by the objective reality of God's self-revelation. The theologian can only begin to develop an understanding and coherent grasp of the Biblical witness "By following through the ontological reference of the various Biblical reports so that his apprehension may be progressively informed and shaped under the self-evidencing force and intrinsic significance of their objective content, i.e. the self-revelation and self-communication of God through Jesus Christ and in the Holy Spirit".[106]

Holy Scripture, the writings of the Prophets and the Apostles, therefore play a basic role in serving and mediating God's self-revelation and self-communication. But it is just because of this that they have also their authority over and in the Church. It is only through indwelling their witness that our minds fall under the self-attesting authority of God's revelation. Consideration of Polanyi's concept of marginal control will further clarify this point.

We have in our introduction looked at the two-levelled structure of a machine, where the parts of a machine, subject to the laws of physics, chemistry and mechanics are nonetheless open to the conditions under which the parts of the machine may be made to operate for a given purpose; they are open to control by the operational principles of engineering, these engineering principles not being definable in terms of the laws of mechanics or the laws of physics or chemistry. This example of the machine was taken as an example of the multi-levelled structure of reality.

We can perhaps apply this concept of a hierarchy of levels to theology and especially in connection with the Apostolic witness. The first level is the human words of the Apostles subject to all the normal rules of grammar and referring to created reality. But the words of the Apostles, because of the Incarnation and Revelation of God in Christ in the power of the Spirit are now opened up to organisation by a higher level of reality, the level of God's self-revelation in human flesh, and are made by reference to Jesus Christ to point beyond created reality to the Creator. Their words are now organised, and given a distinctive shape and form, by this higher "principle" to fulfil a different purpose. The meaning of the Apostles' words cannot now be found by reference to created reality alone, it cannot be found on that level, but only by reference to the revelation of God in the Word made flesh, in the power of the Spirit. We take seriously the words of the Apostles as human words yet they are not subject to explanation on that level. We have to take seriously the Act of God within creaturely existence in the Incarnation. Now the human words of the Apostles only become intelligible and meaningful when correlated with that higher level. Thinking again of the two levels of the operational principles of engineering and the laws of physics in a machine, the principles of engineering by the principle of boundary control imposes upon nature "patterns, artefacts, happenings etc. . . . beyond anything that nature is capable of producing merely in accordance with its own laws".[107] The higher level of the principles of engineering can transform and impose someting new on the laws of nature — and the laws of nature are in no way infringed.

So too the Apostolic witness is something new, it is "inspired", "transformed". Their words are still human words, still subject to the laws of grammar etc., yet transformed by their reference to this new level in the Act of God in Jesus Christ. Furthermore, that Revelation cannot be explained or defined in terms of the human words of the Apostles, no more than can the operational principles of a machine be defined in terms of the laws of physics or chemistry. The parts of a machine are open to organisation by a higher level of reality but do not define that level of reality. So in the same way the human words of the Apostles are open to organisation by a higher level of Reality and can in no way define or explain that Reality. They serve it. Indeed the opposite is the case, that God's self-revelation explains the rise of the Biblical witness, just as the operational principles of a machine give meaning to the parts of a machine. Let us just pursue the analogy of the hierarchy of levels in a machine with the hierarchy of levels in the Apostolic witness a little further. Just as the breakdown of a machine is unexplainable in terms of the operational principles of the machine, and explainable only in terms of the laws of physics and chemistry governing the working of its parts, so too, because the Apostolic witness is a human witness, indeed the witness of sinful men, they may come to attest the divine revelation only very imperfectly. Their failure may hinder our apprehension of God. Their witness has to be attested in the light of that which they attest. Just as we cannot have a machine without its parts, parts that are liable to failure, so neither can we have Divine Revelation without a human witness to that Revelation a human witness that is subject to the possibility of error. But that in no way detracts from the authority of the Biblical witness. We can distinguish quite clearly between the two levels of a machine, the level of its parts subject to the operation of the laws of physics and chemistry, and the level of the machine itself, subject to the operational principles of engineering. We can distinguish between these two levels but we cannot separate them. We cannot have a machine without its parts. So too with the Biblical witness. We can distinguish between the self-witness of Christ and the human witness of the Apostles to that witness, a human witness given shape and form and meaning by Christ's self-revelation. We can distinguish between the two levels but we cannot have one without the other. It is only in the voice of the Apostles that we hear the voice of Jesus.

This brings us to the unique position and role of the Apostles. Only they stand as we have already noted in an immediate relationship to the "direct, absolute, material authority of God, Christ and the Holy Spirit".[108]

They witnessed to the self-witness of Christ and in the Spirit understand and grasp the meaning of that Revelation. By the aid of that same Spirit they were able to articulate their understanding of Revelation, their words were opened up beyond their ordinary creaturely reference to find their true meaning in the Reality of God's self-revelation, in such a way that others hearing their witness, in the Spirit, also are confronted with the Reality of God. "The Apostles are the first, who not only initiated the series (of future churchmen) as a whole, but who must initiate afresh each individual link in it if it is properly to belong to the series. . . . There is therefore, no direct connection of the Church with Jesus Christ and no direct life by His Spirit — or rather, the direct connection of the Church with Jesus Christ and its direct life by His Spirit is that it should be built on the foundation which He Himself laid by the institution and calling of His witnesses."[109] The Apostles therefore initiate a tradition. Those who accept their witness of God's self-revelation, their understanding and interpretation of events, as their ultimate beliefs enter into the Church, the community of people holding these same ultimate beliefs, passed on from the Apostles.

Polanyi's analysis of the role and position of the great men of science in forming the scientific tradition might helpfully be referred to here. The past masters, the great men of science, are those who have shown great creative originality, "a gift possessed by a small minority".[110] Through their genius they have discovered insights into the understanding of reality, of this intelligible world we live in. In so doing they have initiated a tradition descending from them of those who accept their findings, who are persuaded by them, who trust themselves to these masters and follow them. The records of the works and discoveries of the great men of science are embodied in the articulate framework of the scientific tradition. We look up to these past masters and dedicate ourselves to the discoveries, the standards, the knowledge and ultimate beliefs they have formulated. The whole process of our acquiring scientific knowledge requires our submission to their thoughts, works, and ideals. We accept their authority and in this way the scientific community is informed by one tradition, a scientific community holding the same set of ultimate beliefs, sharing the same practices and values. Only in this way can there be any meaningful dialogue between scientists if they are informed by the one tradition, adhering to the same past masters of science and the values and practices nurtured by them.

Science as we have shown in our introduction, exists and continues to exist, only because its premisses and ultimate beliefs can be embodied in a tradition which is fostered and renewed and checked by the community of scientists, a community which is brought into being by the individual's love of science and belief in science and his devotion to scientific standards. "By apprenticing himself to an intellectual process based on a certain set of ultimates, this newcomer enlists as a member of the community holding these ultimate beliefs and his commitment to these necessarily involves the acceptance of the rules of conduct indispensable to their cultivation."[111] In other words, the acceptance of a definite set of principles handed down from the past masters of science brings forth the scientific community governed by these principles, and the scientific community would cease to exist the moment these principles were cast aside. Through apprenticing ourselves to these past masters and to the intellectual tradition descending from them we come into contact with the same transcendant reality which they first discerned and we seek to discover further aspects of that reality and so develop the tradition. Polanyi also goes on to make the further point, in a slightly different sphere that, for example, "the admirer of Napoleon does not judge him by independent previously established standards, but accepts, on the contrary, the figure of Napoleon as a standard for judging himself. Such an admirer of Napoleon may be mistaken in the choice of his hero, but his relation to greatness is correct."[112] In other words, when we accept the standards and insights of the past masters of science we do so not because we judge them by some previously established standards. No, their standards become the standards by which we judge all else. Our acceptance of the insights and standards of the past masters of science is an a-critical act on our part. We are faced here with an ultimate commitment on our part as to which masters we will apprentice ourselves and try to understand and imitate. No authority can teach us how to make this choice. But after the choice is made we judge all else in the light of the standards we have accepted, we cannot judge them from any higher level.

So with the Apostles, though unlike the past masters of science we must not speak of their discoveries but rather of God's revelation. They stand in immediate relationship to the self-revelation of God in Jesus Christ. We must trust them, accept their authority, their witness and understanding and judge all else in the light of their witness. We do not judge them by our own standards or by any other. Their witness becomes our standard and lays certain constraints on the Church in the same way in which the self-set standards of the scientific community lays certain constraints on them. We either accept the witness of the Apostles and so enter into the community of the Church; the community of all those who accept their witness and through it the self-witness of God, or we reject it. The Apostolic witness becomes a standard for judging the Church. The Church does not judge them by any previously established standards.

We will develop this understanding of tradition and bring out further important aspect of it a little further on when we consider the relation of tradition to the authority of the Church. It is to this latter theme that we now turn our attention.

CHAPTER III

THE AUTHORITY OF THE CHURCH

WE look now at the authority of the Church itself. For this section it might be helpful first of all to compare and contrast Karl Barth's account of the authority of the Church with Polanyi's account of the authority of the scientific community.

Karl Barth makes it quite clear that, though the ultimate authority in the Church is the self-witness of God's revelation in Jesus Christ, nonetheless, the Church has a genuine authority of its own, but it is an authority as he describes it "Under the Word", and therefore under Holy Scripture. The Church has an authority which is established and defined by this prior and ultimate authority of God's self-revelation. It has an authority which is "mediate" and which Karl Barth illustrates by reference to the sun's reflection in water.[113] We see in the Church a reflection, but only a reflection, of the "immediate" authority of Christ. The Church has an authority only in so far as it acts as a secondary witness to the primary witness of the Apostles, only as it seeks to demonstrate the revelation of God in Christ by its whole life.

As we look at this authority of the Church, and also at the authority of the scientific community, it might be helpful at the outset to distinguish between two different aspect of that authority. We have to distinguish first of all between the authority which the Church has over say the layman or the catechumen, and secondly the way in which that authority is established in the Church itself. This will be seen to be similar to the structure of the scientific community. There is first of all the authority of scientific opinion over the student, an authority which establishes what is scientific and what is not, and secondly the recognition that the authority of scientific opinion within the scientific community itself, is a mutual authority established between scientists and not above them.

With regard to the authority of the Church Karl Barth notes first of all that we ourselves cannot make our own confession of faith unless we have first heard the witness of the Church. How else can we come to believe unless we hear what we have come to believe first of all from those who already believe? We depend on the prior witness of the Church. "For that very reason I recognise an authority, a superiority in the Church: namely, that the confession of others who were before me in the Church and are beside me in the Church is superior to my confession. . . ."[114] If we are ever to come to understand the Christian faith we must listen to those who already believe, we have to trust and respect and accept the witness of the members of the Church and so recognise that it is for us an authority. This is very similar to how Polanyi describes the situation of a student of science. The student must accept the authority of the scientific community as to what "science" is science. He must listen to them, trust them so that eventually through the teaching he receives and through trusting himself to the scientific tradition, he apprehends what is the objective content of their teaching. Before a student of science is in a position to be confronted with the direct ultimate authority of the transcendant reality embodied in the scientific tradition he must first receive his knowledge of it indirectly through the scientific community. To that extent the scientific community has an authority which the layman and the student must accept and respect. ". . . the process of learning must rely in the main on the acceptance of authority. Where necessary this acceptance must be enforced by discipline. Naturally, there is a field of possible conflicts between masters and pupils. The student who, on obtaining in the course of elementary practice an erroneous result from his chemical analysis, would claim to have made a fundamental discovery, would make no progress. He must be reprimanded and if necessary removed."[115]

This is similar to Barth's description of the authority of the Church when he says "my first duty is to love and respect it (the confession of the Church) as the witness of my fathers and brethren. And as I do so, as I recognise the superiority of the Church before and beside me, it is to me an authority."[116] He makes it clear however that this authority is an authority "under the Word". "Before both and therefore above both (the confession of others and myself) is the Lord of the Church with His Word. Only under His Word can some confess and others hear their confession before they confess themselves. Under His Word there is therefore a genuine authority of the Church."[117] The authority of the Church is "mediate" and indirect and meant to point us to the direct and immediate authority of God's self-revelation. We will return to this particular point a little further on.

This prior witness of the Church to which we listen, and more especially the catechumen has to listen, and which has an authority under the Word, reaches us in the form of the written confessions and the dogmas of the Church. How these confessions are arrived at and by whom Karl Barth describes in this way. There is a constant enquiry or debate in the Church into the faith of the Church to ensure that the Apostolic witness has been truly heard and understood and received. To this end each member of the Church must make known to others in the Church his understanding of the faith he has received. "I do not do this to force it on the Church in the peculiar form in which I necessarily hold it, as though I were presuming either to want to or be able to rule in the Church with my faith as it is mine. On the contrary I do it to submit to the verdict of the Church, to enter into debate with the rest of the Church about the common faith of the Church, a debate in which I may have to be guided, or even opposed and certainly corrected."[118] We enter into a debate in which we ourselves have to be open to the judgement of others and in which they also are open to what we have to say to them. This debate takes place under the objective reality of the inherent rationality of the divine revelation, "under the

Word". Each side of the debate is trying to point the other to the understanding of that revelation they have grasped as well as critically judging their own understanding in the light of what others have understood of that same reality. In this way our understanding is tested in the light of that to which it refers.[119]

In this way the authority of the Church rests on a "decision",[120] reached by the Church where out of the debate and enquiry concerning a true faith agreement is reached. This whole procedure is illustrated by Karl Barth with reference to the formation of the canon of Holy Scripture. As previously mentioned when we consider the doctrine of Holy Scripture we enter into the logical circle of self-asserting, self-authenticating truth. The Scriptures speak for themselves with divine authority. The Church can only confirm that here in these writings it has heard the Word of God. The establishing by the Church of a canon of Holy Scripture is its confession, its counter-witness to the self-witness of Scripture itself. Those entering the Church must accept on the authority of those before them in the Church that these writings are, and others are not, the canon of Holy Scripture. The authority of the Church is normative. The newcomer, the catechumen accepts the decision reached by the Church and in these writings accepted by the Church seeks to hear what the Church has heard. The catechumen must accept the canon of Holy Scripture and there seek to find the Self-witness of God's revelation in Christ. If that Self-witness is found in only parts of the canon, nonetheless, the catechumen must trust the decision reached by the Church and believe that as he goes on he will hear the Word of God in all the rest as well, as have others before him. It is also of the utmost importance that the decision reached by the Church as to what constitutes and what does not constitute the canon of Holy Scripture be a unanimous decision. There must be agreement. "There must obviously be a chorus or choruses (of fathers and brethren) and not a confusion of many independent voices."[121]

The Church exercises authority over the layman in the

same way in which the scientific community exercises the authority of scientific opinion over the student. But within the Church there is a common enquiry into the faith of the Church. This particular point is brought out clearly by Polanyi with reference to the scientific community. As we have previously indicated a student must accept the authority of the scientific community and its traditions if he would ever hope to understand science. This authority of the scientific community is concretely seen in the natural laws of science. Indeed we can go so far as to compare the natural laws of science with the creeds and dogmas of the Church. The natural laws of science are "open" laws. They are open to the transcendant reality of the inherent rationality of the world and they point beyond themselves to that reality. They arise as the result of a decision made by the scientific community as a whole to accept one of a number of possible interpretations that might be placed on experimental data as being the one that truly intimates the rationality of the real world. As we have already noted in our introduction, in times of scientific controversy several coherent systems of interpretations can be placed on the available experimental data. For example this was true in the case of the controversy between Pasteur, Liebig and Wohler over the nature of alcoholic fermentation.[122] Whenever two such interpretations confront each other, controversy ensues. There takes place a process of persuasion, debate and argument, sometimes severe, until one or other of the interpretations gradually gains acceptance by the scientific community as the one which reflects in its own rationality, the inherent rationality of the world. This is a "decision" responsibly reached by the community of scientists. Even apart from times of scientific controversy the scientific community still advances its knowledge by the principle of "overlapping neighbourhoods" where each scientist judges the work of his neighbour and his own work in turn is judged by his fellow scientists. Each scientist questions and listens to the other. There is a common enquiry into the truth.

This "decision", this scientific opinion which forms the authority of the scientific community, and which results in the formation of the natural laws of science, must be respected by the student of science. But Polanyi makes the important point that this authority is a "competent" authority but not a "supreme" authority.[123] In accepting the authority of scientific opinion we must always remember that this opinion is the result of a value-judgement made by the scientific community. There is therefore always present the possibility that it could be mistaken. Nonetheless, having said that, the decisions of the scientific community are responsibly reached and therefore competent and so the student must accept on the authority of the scientific community the natural laws, the doctrine, formulated by it, and indwell it as a "clue" in reaching an heuristic vision of reality. As we have already mentioned the natural laws of science are "open" laws. The student must attend from the doctrines of science to the reality described by science and so gain an intuition of that reality. In this way the student is brought into direct contact with the reality of nature and the authority of the scientific community fulfills its purpose and is to some measure eclipsed. "As he approaches maturity the student will rely for his beliefs less and less on authority and more and more on his own judgement. His own intuition and conscience will take over responsibility in the measure in which authority is eclipsed. This does not mean that he will rely no more on the report of other scientists — far from it — but it means that such reliance will henceforth be subject to his own judgement. Submission to authority will henceforth form merely a part of the process of discovery, for which — as for the process as a whole — he will assume full responsibility before his own conscience".[124] Thus it is that the authority of the scientific community is a competent authority but not a supreme authority. It brings the student into contact with reality but in the light of that reality the mature student may then call in question the tradition, the doctrine of science. The authority of the scientific community is not a "supreme" authority. The supreme authority rests with the intrinsic intelligibility of objective reality, in its own self-attesting truth.

We can compare this "competent" authority of the scientific community with how Karl Barth describes the authority of the Church as being an authority "under the Word". Using Polanyi's language we can say that the authority of the Church is a competent authority but not a supreme authority. The dogmas and confessions of the Church are "open", they are meant to be indwelt as clues and integrated to form an heuristic vision of God. We are meant to attend from them to God's own Self-revelation and in the light of God's own Self-revelation perhaps correct or call in question these same dogmas and confessions. Indeed we cannot help but note how similar Karl Barth's description of the authority and function of the Church's confessions is to Polanyi's description of the authority and function of scientific opinion. The language he uses is different but the intention is the same. Barth maintains that "it is not by agreeing with these statements and appropriating them, but by learning the direction from these statements, that we respect the authority of the confessions. For that reason it may well be that in learning the direction from it we have to oppose critically certain or even many of its statements".[125] Again he says, "We cannot allow ourselves to be bound by the confession as by law. Does that mean an end of its authority and respect for it? Not at all — we might say rather that it only begins at this point. For at this point it takes a third and genuinely spiritual form. It becomes a constant antithesis, the horizon of our own thinking and speaking. Naturally within this horizon it is a question of our own free thinking and speaking, for which we must bear the responsibility, which is not bound by any law, except that of its object and therefore only by Holy Scripture. Our study of the confession, our hearing of it as leader and key witness, is now behind. We must now take the Word itself . . .".[126] We follow the direction of the confession, we attend from it to the Word itself, in the light of which we may call the confession in question. In that sense the authority of the

Church and of its confession and dogma are a competent authority. Though they have been responsibly reached, they may yet be in need of correction. Nonetheless we must accept their authority, we must make them the horizon of our thinking, the subsidiary clues we indwell, and follow their direction.

Illustrating this by the example of the formation of the canon of Holy Scripture, we can say that the catechumen must accept the authority of the Church as a competent authority as to what does and what does not form the canon of Holy Scripture. This means as Karl Barth points out that the canon of Holy Scripture is not "closed", but "open", and the possibility of a new canon is not excluded.[127] As the catechumen matures in the Christian faith and becomes better informed he may wish to call in question the decision reached by the Church. This is seen for example in the Reformed Church's rejection of the Apocrypha, and also in Luther's concern over other parts of the canon. But those who call in question the decision of the Church, or its need of correction must seek to convince the rest of the Church in the light of Holy Scripture and Holy Spirit that a new decision has to be taken. The Church must always examine its confession and confirm it or make a new and better one.[128] Holy Scripture is the supreme authority.

The scientific community upholds a "competent" authority over the lay public and controls as a result the process by which students are trained to become members of it. But once the student has undergone the critical discipline of authority and has become an independent scientist, once this authority has served its purpose and brought the student into contact with the reality of nature, the student's own conscience will take over responsibility for the developing of his own beliefs about that reality. There is no longer any superior authority of his fellow scientists "above" him. Yet this does not mean that the scientist will from then on be only subject to his own judgement. "His submission to scientific opinion is entailed now in his joining a chain of mutual appreciations, within which he is called upon to bear an equal share of responsibility for the authority to which he submits".[129] As mentioned above,[130] and in the introduction,[131] each scientist questions the work of his fellow scientist in the light of the intrinsic intelligibility of objective reality into which they enquire. It is only in this way that science can progress. If each scientist were independent in the sense of cutting himself off from the findings and criticisms of other scientists, progress in science would stagnate.[132] Here we have now for the independent scientist, as contrasted with the student, a *"mutual"* authority *"between"* scientists, which leads on to what Polanyi calls the "general authority" as opposed to the "specific authority" of science.[133] "Every time a scientist makes a decision in which he ultimately relies on his own conscience or personal beliefs, he shapes the substance of science or the order of scientific life as one of its sovereign rulers. The powers thus exercised may sharply affect the interests of his fellow scientists. Yet there is no need for a paramount supreme power to arbitrate in the last resort between those individual decisions. There are divisions among scientists, sometimes sharp and passionate, but both contestants remain agreed that scientific opinion will ultimately decide right; and they are satisfied to appeal to it as their ultimate arbiter."[134] The authority of the scientific community, the competent authority which it exercises over the layman is a "general" authority. No one man or one group, nor any specific authority imposes on the rest the decision as to what "science" is science. This would be destructive to the progress of science.[135] The decisions are reached by the scientific community as a whole. Behind this lies Polanyi's belief, and the scientist's belief that each individual can make genuine contact with the reality to which the doctrine of science points. Each scientist must judge his own thinking and the thoughts of his fellow scientists in the light of that reality, and of what others have to say of it.

We have already looked at how Karl Barth has described the way in which the confessions of the Church are reached through debate in the Church, and how his account of the process in the Church is similar to that of Polanyi's account of what takes place in the scientific community. Indeed Polanyi himself declares that the structures of both are similar and turns to an examination of authority and tradition in the Church to help shed some light on the understanding of tradition and authority in science. Again using Polanyi's language we can say that the "competent" authority which the Church exercises over the layman, is an authority that is a "general authority". There can be no possibility of one man or one group within the Church imposing decisions on all the rest. We will develop this point further when we come to discuss the differences between the Reformed and Roman Catholic understanding of tradition and authority in the Church.

Before we return again to look a little more closely at tradition and its interrelation with authority, which we have now briefly looked at, let us first of all enlarge on two further points in connection with the authority of the scientific community and the authority of the Church. Karl Barth points out that the confessions made by the Church are not a comprehensive exposition of the Christian faith. They have always arisen to counter a particular error and so their material content is limited in that way. When the existing confession of the Church, commonly held by the Church, is called in question, or when some new circumstances arise which give rise to different teaching on the basis of the same confession, the unity of the Church is thus obscured, and there arises the need for a "new" confession which will in the light of the new circumstances express the unity of the Church's faith clearly once again.[136] This new confession has as always to be judged in the light of Holy Scripture but nonetheless it has arisen in response to error and confusion and is therefore limited in that way. It is limited by the circumstances out of which it arose. It is not a comprehensive insight in God's Self-revelation. And as Karl Barth goes on to point out, in addition to this material limitation it also has geographical and temporal limitation.[137]

This leads him to ask the question how then the

Church's confession can have any authority at all. That question he goes on to answer by pointing out that the confession of the Church cannot be defined in terms of its limitations. They limit the confession, yes, but they do not determine it. The confession is determined "from above" and not "from below". As we mentioned above the Church's confession must become the horizon of our own thinking and speaking, it must direct us to God's Self-revelation. "Primarily and decisively it is not the wording of the Church's confession, nor its form as geographically and temporally and historically limited, but its direction whose exposition can constitute the necessary horizon of the present day Church and as such have and be authority."[138] Again he says, "It is not by agreeing with these statements and appropriating them, but by learning the direction from these statements that we respect the authority of the confession".[139] It is in this way that the confession of the Church is determined "from above" and not "from below", from its temporal historical and geographical limitations. That is why as we follow the direction of the confession and come to the Reality which gave rise to it, the confession itself may demand its own revision in order to point to that Reality. The confession therefore does not bind us to itself but to the self-witness of Holy Scripture to which it points. In that way "No obligation to a confession can ever be anything but an obligation to Holy Scripture".[140]

Polanyi also finds similar limitations attached to our acceptance of the traditional framework of scientific thought and he raises a similar question in regard to science as Barth raises in regard to theology, and gives basically the same answer to it. He asks, "How can we claim to arrive at a responsible judgement with universal intent, if the conceptual framework in which we operate is borrowed from local culture and our motives are mixed up with the forces holding on to social privilege? From the point of view of critical philosophy this fact would reduce all our convictions to the mere product of a particular location and interest."[141] In other words the framework of scientific thought is limited geographically and temporally in much the same way in which the Church confessions are limited. Polanyi answers the questions in the following way. He begins by reminding us of the fact that the operational principles of a machine would always guarantee the successful working of the machine but for the fact that the machine is composed of parts which are liable to break down and so cause the failure of the proper functioning of the machine. This we have already fully discussed. In the same way a human being would be able always to make responsible decisions but for the fact that such responsibility can only be exercised by responsible control of all our self-centred drives. Such innate self-centred drives may lead us into failure to act responsibly. Again on a different level all our mental achievements rely on the proper working of all other parts of the body, which therefore limits the proper functioning of our mental powers. "Everywhere the potential operations of a higher level are actualised by their embodiment in lower levels which makes them liable to failure."[142] In all the above examples the lower level sets certain limitations to the performance of the higher level, and yet at the same time provides the opportunity for the operation of these higher principles. So with our acceptance of the articulate framework of science. We must submit ourselves to these limitations which are beyond our responsibility and accept them and dwell in that framework. "I accept the accidents of personal existence as the concrete opportunities for exercising our personal responsibilities. This acceptance is the sense of my calling."[143] We must accept these limitations and see in them opportunities for exercising our responsibility for pure thought. In other words the scientific framework is determined by its reference to a transcendant reality "above" it even though it is limited from below. Within the particular limitations of our time or place we make decisions responsibly with universal intent, which may call for the correction of our particular framework of thought.

The second point we must bring out more clearly before

we go on to look at the interrelation of tradition and authority is to make it quite clear that the authority of the Church does not imply any kind of authoritarianism on the one hand nor, since for example each individual can question say the composition of the canon of Holy Scripture, any kind of self-will or subjectivism on the other. Just as the Church has an authority under the Word so does it also have freedom "under the Word".[144] This freedom "under the Word", as Karl Barth describes it, rules out any idea of authoritarianism in the Church or any thought of arbitrary self-assertion. The antithesis between authority and freedom is completely overcome when we accept the reality of a Word of God for the Church and our free obedience to that Word. Michael Polanyi's philosophy overcomes just such a false antithesis. His concept of personal responsibility in science rules out both authoritarianism and subjectivism. Objective reality confronts us, thrusts itself upon us and yet we are free to turn away from ourselves to that objective reality. We are free to say yes to what is there. It is in such a way that we become rational beings, behaving in accordance with what is there and its inherent rationality. In the same way the Word of God confronts us and calls forth our obedience or our disobedience. There is freedom under the Word. One aspect of this freedom in the Church Karl Barth illustrates with the interpretation and application of Holy Scripture. Each individual in the Church must be willing himself to assume responsibility for the interpretation and application of Holy Scripture. It is only in this way that the "debate" within the Church, previously discussed can possibly arise. There is this freedom in the Church. "The founding, maintaining and governing of the Church by Scripture does not happen in such a way that the members of the Church are only spectators or even objects of this happening. They rather become subjects of it."[145] The whole Church has the task of interpreting Holy Scripture. To deny this responsibility which is laid upon every member is to exchange the "general authority" of the Church for a "specific authority", to deny freedom to each member to make his own response to the Word and indeed to imply that members cannot themselves be confronted directly with the Reality of God in Jesus Christ. This would lead to authoritarianism in the Church. On the other hand, it must be remembered that this freedom is defined and delimited by reference to the Word of God. It always confronts us and calls forth our obedience to it. This rules out any accusation of subjectivity or self-will. Again to use Polanyi's language the Christian's beliefs are "personal", his knowledge of God is a personal knowledge in the very same sense in which Polanyi maintains that the scientist's knowledge is a personal knowledge. Our faith and knowledge arise as our rational response to and acceptance of what is objectively given, when we turn away from ourselves to that Reality confronting us.

Karl Barth makes the further point that the freedom of interpretation in the Church arises only because the Word of God is given us "in concerto", in the words of the prophets and Apostles and not "in abstracto". He points out that the Word of God itself is clear and in no need of explanation, but because we receive it in human form in the words of the prophets and Apostles this means that their words like all human words are capable of being understood or misunderstood by us. Hence the need of interpretation and explanation. Furthermore he adds that "What makes the Word of God, in the form in which we encounter it, obscure and in need of interpretation are the ideas, thoughts and convictions which man always and everywhere brings to this Word from his own resources".[146] In addition to our need to make sure we understand aright the human speech of the Apostles we have also to be careful that we do not read into their witness our own ideas and thoughts so making obscure the word of God, which as already mentioned is in itself quite clear. It is because of this that there arises the need for interpretation and explanation. This responsible interpretation of Holy Scripture is the task of each member of the Church as we seek to follow the direction of the Apostolic witness. We have this freedom and exercise it responsibly under the Word.

Chapter IV

THE INTERRELATION OF AUTHORITY AND TRADITION

WE turn now to look again at the role of tradition in science and theology and, now that we have looked at the role and nature of authority, to look at the interconnection and interrelation between authority and tradition. Polanyi notes this interrelation when he says "The authority of science is essentially traditional".[147] When we looked at the unique position of the past masters of science we noted that it was they who initiated the tradition of science. Their values and practices and beliefs were transmitted to each succeeding generation in their writings and works and each new member of the scientific community had to devote himself to these same standards and uphold them. We recall too that these standards are self-set standards and ones that are at first accepted a-critically as the student of science "apprentices himself to an intellectual process based on a certain set of ultimates".[148] All members of the scientific community hold the same standards and beliefs which in turn means that they accept the same group of persons as masters. It is this fact which explains how it comes to be that despite the fact that each scientist follows his own personal judgement in believing any one particular claim to be true, and despite the fact that each scientist chooses his own problems and pursues them and verifies his own results according to his own personal judgement, there is nonetheless a consensus among scientists as to what is "unscientific" and what is "scientific". "Even though controversy never ceases among them, there is hardly a question on which they do not agree after a few years discussion".[149] This fundamental agreement among scientists which exists despite the individualism of each scientist is illustrated as Polanyi points out by the very fact that each scientist tries to persuade his fellow scientist of the truth of his claims.[150] The scientist believes that his fellow scientist will come to recognise, eventually, his claims and agreement be reached. This belief is based on the conviction that all scientists operate from the same premisses, with the same ultimate beliefs and standards and confronted by the same reality. This being the case, that each accepts the same traditional premisses of science, it is inevitable that "however far we may advance thence by our own efforts, our progress will always remain restricted to a limited set of conclusions which is accessible from our original premisses".[151] Because scientists operate with the same ultimate beliefs and values and premisses, agreement between them is possible. The eventual harmony among scientists, the general authority of science, is achieved because every scientist is informed by the same tradition deriving from the same set of past masters. "The origin of the spontaneous coherence prevailing among scientists is thus becoming clear. They are speaking with one voice because they are informed by the same tradition. We can see here the wider relationship, upholding and transmitting the premisses of science, of which the master-pupil relationship forms one facet. It consists in the whole system of scientific life rooted in a common tradition. Here is the ground on which the premisses of science are established; they are embodied in a tradition, the tradition of science. The continued existence of science is an expression of the fact that the scientists are agreed in accepting one tradition, and that all trust each other to be informed by this tradition."[152] Yet Polanyi makes it quite clear that it is not to the person of these past masters that we submit ourselves, but rather is it the case that we submit ourselves to what we understand to be their teaching. In the light of the ideals and standards formulated by these past masters, self-set standards a-critically accepted by us, we may challenge some of their works as being not compatible with these same ideals.

The General authority of the scientific community rests

therefore on the fact that each member of the scientific community is informed by one and the same tradition, that each accepts the authority of the same set of past masters and the ideals embedded in their works (ideals and standards tacitly handed down from each generation). In this way the scientist can appeal from the tradition as it is to the tradition as he feels it ought to be, and because all scientists share the same tradition the grounds of agreement between them are laid.

All of this is obviously of relevance to our understanding of the confession of the Church which is the outcome of a decision reached by the whole Church. The unity of the Church, and the authority of its confession, is maintained only when each member of the Church accepts the authority of the Apostolic Witness and the ultimate beliefs laid down by them. If the members of the Church accept and recognise any other authority alongside of and equal to the authority of the Apostles then confusion will arise. No longer will they be starting from the same premisses. No meaningful dialogue can take place. No agreement can be reached. The Church is Apostolic, and therefore one, grounded upon an unrepeatable foundation, a foundation laid once and for all. It is only as the Church is directed back to the Apostolic witness and builds upon it that it becomes the Church. There can be no other authority, no other foundation laid in the Church. Just as the scientific community is established as each member devotes himself to the same set of masters, practices and standards, so the Church too is established as each member devotes himself to the witness of the Apostles and accepts them as his authority, and their standards and ultimate beliefs the standards by which all else is judged. Indeed the Apostolic witness can be compared with the self-set standards of plausibility, scientific value and originality. These self-set standards of the scientific community are the means by which all that is trivial and unscientific is kept out from the body of knowledge accepted as science. This we have already looked at in the introduction.[153] Acceptance of these standards is what constitutes the scientific community and lays at the same time certain constraints on it. So too the Apostolic witness is the "self-set standard" of the Church, a standard that also lays constraints on the Church. The dogma of the Church can also be regarded in the very same way.

The "general authority" of the Church, the common confession of the Church, rests therefore on the fact that those taking part in the debate are informed by the Apostolic witness accepting it as their norm. Agreement will then be possible in the Church because each person is starting from the same premisses. There will be one Church, one confession, one Holy Catholic and Apostolic Church. In this connection we again remember as Karl Barth points out that it is not to the person of the Apostles that we submit ourselves but to the voice of Jesus heard in the voice of the Apostles. This may lead us to call in question certain aspects of the Apostolic witness in the light of that to which the Apostles refer in the same way in which we are critical of some aspects of the works of the past masters of science in the light of the ideals which they themselves have put forward.

This understanding of tradition therefore helps us to understand a little more clearly the difference between the general authority and the specific authority discussed previously. The general authority of science is one that lays down and fosters certain premisses and ultimate beliefs, a set of general presuppositions embodied in a tradition. Through his acceptance of and devotion to these presuppositions each individual is brought into contact with the "transcendant spiritual reality" embodied in the tradition and transcending it. He is brought into contact with that reality in the light of which he can judge and correct the tradition. Thus we have a consensus of separate individuals all of whom are rooted in a common tradition. A specific authority, however, "imposes conclusions",[154] and assumes that only those exercising this specific authority have contact with the source from which all the tradition springs. "We see emerging here two entirely different conceptions of authority, one demanding freedom where the other demands obedience".[155]

We can perhaps at this point examine the role of the teachers and Councils of the Church and again draw a comparison with Polanyi's account of the way in which the premisses of science are fostered and guarded and developed by what he calls the set of "institutions" within the scientific community. Although the scientific community is self-governing, the authority of science resting with scientific opinion at large, nonetheless authority is not equally distributed among scientists. The opinions of certain scientists, certain competent recognised experts in their fields of study, carry more authority and are more highly valued than the opinions of others. Their opinion focuses and expresses the opinion of the scientific community as a whole. Such scientists have responsibility for teaching students, for the running of scientific establishments, for appointments to scientific posts, grants, research, selecting suitable contributions to journals and rejecting others, approving and recommending books. Their task is a regulative one. Their task is to maintain and foster the standards and premisses of science embodied in the scientific tradition deriving from the past masters of science. They do not create these premisses, they maintain them, they guide the progress of the scientific community which is established on these premisses. Their function is protective and regulative. This functioning of the set of scientific "institutions" can be compared to the functioning of the teachers and Councils of the Church. Let us look, by way of illustration, at the position of Calvin and Luther in the Reformed Churches. They are both recognised authorities in the Church and indeed have a genuine authority in the Church. "Neither in principle nor in practice, therefore, can we deny the existence of the ecclesiastical authority of specific teachers in the Church. But if this is the case, then it is of itself understandable theology — assuming that it is a fact — that in the Evangelical Churches it was the Reformers who acquired this authority. If our Churches confessed that they were reformed by the Word of God and not simply by Luther and Calvin, their reformation did take place by the witness borne to them by Luther and Calvin. Therefore the witness of Luther and Calvin is decisive and essential for their existence as this Church, as the Churches reformed in this way, and therefore for the whole contingency of their existence as the Church of Jesus Christ. This may not be true as a constitutive, but it is certainly true as a regulative principle. If they free themselves from this witness they are no longer these Churches and therefore no longer contingently the Church of Jesus Christ."[156] Their task is a regulative one. The ultimate authority still rests with the Apostolic witness. The authority of the teachers of the Church and of the Councils of the Church are recognised only in so far as they foster and maintain the ultimate beliefs handed down from the Apostles. So it is that the Church must listen to their teaching and to the teaching of the Councils of the Church and must always declare whether or not this teaching itself serves to help others understand the Word of God. Each teacher must be assessed by the standards of the Apostles. Only as they point us to the Apostles do they speak to and for the Church. In the same way as the institutions of science voice the opinions of the scientific community as a whole so too do the voices of the Councils and the teachers of the Church voice the opinions of the Church as a whole. They speak to and for the Church. They do not create the premisses on which the Church is founded, they merely guard these premisses and call the Church continually back to them.

Another aspect of tradition which we must deal with further is one of the points made previously in the introduction — the fact that there are formal and informal elements in every tradition. We have seen that the act of scientific discovery is a skill similar to the skill of perception, a skill that cannot be explicitly defined. We have seen too how we cannot make completely explicit our scientific knowledge, "We know more that we can tell". So we cannot explicitly lay down the premisses of science as if they could be formally passed on from generation to generation in a detached impersonal way. In other words we must, for example, learn the skills of scientific research from imitating the skill of a master of scientific research, and so receive subsidiarily the premisses of science which we can then later, by analysis of the skill we have learned, know focally. The student too must learn the proper integrative skill which will enable him to indwell as subsidiary clues, the dogma of science, its natural laws, and so grasp that reality to which it refers. This integration of the subsidiary to the focus is always the act of a person.[157] This is a skill he has to learn by accepting the authority of his teacher, even if in fact the teaching he receives at first appears meaningless to him. He must believe in science, accept and trust the authority of his teacher and seek to discover the meaning of that teaching by achieving a similar kind of indwelling as the teacher is practising. Hence the reason why, as previously mentioned, there is little progress made in science in non-European or non-American countries where there is no scientific tradition, where scientific skills and the premisses of science cannot be tacitly learned from an authoritative representative of that tradition. This aspect of tradition is further illustrated by Polanyi with reference to the way in which a child learns. A child learns to speak by assuming that the words that their parents speak in their presence are meaningful. The child has to learn what we mean by our words and attend from our words to that to which they refer. There can be no explicit way of conveying an understanding of the meaning of the words we use. "To know a language is an art, carried on by tacit judgement and the practice of unspecifiable skills. . . . The tacit coefficients of speech are transmitted by inarticulate communications, passed from an authoritative person to a trusting pupil, and the power of speech to convey communication depends on the effectiveness of this mimetic transmission."[158] Thus begins the whole process by which the child begins to indwell the articulate framework of society. He begins by accepting the authority of his parents that what they say and do is meaningful. By imitation, by practice, by a whole series of tacit judgements the child achieves a similar indwelling in language as do his parents and so grasp the meaning of words and of language. As he grows older he will in the same way accept the authority of the intellectual leaders of his society, assuming again that what they say and teach is meaningful. From childhood to manhood he must believe before he can understand. There must be a traditional framework of thought, where the tacit skills of learning a language, or of acquiring a knowledge of science, are passed on from master to apprentice, teacher to student, parent to child. A traditional framework of thought cannot be made wholly explicit. It contains tacit elements that can only be informally passed on.

All of this is obviously of great importance for our understanding of Christian Education. There must be in the Church practising Christians who have achieved an integrative indwelling of the Biblical witness, who indwell that witness to form a heuristic vision of God under the impact of the self-revelation of God through Jesus Christ. Only they can pass on that skill to the newer members of the Church. The formal elements of Christian Education need the informal.

But in addition to the relevance of Polanyi's understanding of tradition to adult Christian Education we must note also its special relevance to the Christian education and upbringing of children. By the age of four it is estimated that about 40% of the mental development of a child has taken place. By that age the child has learned as much as he will in the next ten years of schooling.[159] Indeed much of what he later learns will only be making known focally, by analysis, the premisses of the skills he has tacitly acquired in his early years of infancy. This underlines the great importance of the home in the child's Christian upbringing especially its influence on young infants. "Before we begin actual bible teaching, we have been inculcating the basic view in our daily routine. Every event, every experience constitutes a 'learning situation' for the children. Inevitably, as the years pass, a response to life is being established. Recent research has shown that children develop this 'response to events' quite early. If these events are interpreted through the parents' pattern of belief, the child can absorb them unemotionally. Since the parents' believe that God is in command, that He is a loving Father, it follows that all things work together for good for those who love Him. We teach, therefore, to look for the Father's love and concern in any event however puzzling — in the successes and joys; but also in the seeming failures or disappointments — that even a crucifixion will have a resurrection. This positive attitude to life can be taught best by the parents in the day to day problems which occur — it is not something read about, nor taught in theory by a Sunday School teacher — it is experienced in actual life — it becomes a family pattern."[160] In other words the child needs a framework of thought which will give meaning to all the various events of life. That the child learns from his parents, from their pattern of behaviour. The child assumes that the parents' response to various situations are meaningful; it will seek to achieve a similar response, and so the child acquires the parents' pattern of behaviour. As the child seeks to assimilate the parents' pattern and framework of thought he receives subsidiarily the premisses of the Christian faith which can then later be known focally. If the parents do not believe and do not show a Christian response to, and understanding of, all the events of life, if the Church, prayer and worship are not important to them and part of their pattern of life they will not be for the child. If these things mean nothing to the parents, they will mean nothing to the child. They will not be regarded as being important for the understanding of life or the various events of life. In that case a different pattern of family life will be established and assimilated and so the child will not tacitly acquire a knowledge of the love of God which can later be made explicit.

But we must also note that the whole Church, the whole Christian community has a role to play in the child's education. "The foundations of Christian education are laid in the home, but 'basic in this education is sharing in and being moulded by the fellowship of those who have accepted the Gospel'. For good or ill, therefore everyone in the Church is a Sunday School teacher. The children come into the fellowship as learners. They have heard about prayer and praise at home, but they need also to experience what that means when mother and father join in worship with others in Church. There is a totally new dimension here which cannot be taught or experienced outside the fellowship."[161]

We might here perhaps draw another parallel between the way in which Polanyi describes the development of the child in society with its development in Church. As already stated the child learns the meaning of a language from its parents in the home. But embedded in that language and the actual framework of thought which it indwells are the beliefs, the values, the standards of society as a whole. For example in Western Society our language has embedded in it the premisses of a naturalistic outlook on the world as opposed to say the magical one. If a child wants to understand further the premisses embedded in an articulate framework of thought he has to move out of the home, where the tacit foundations are laid and grow into the scientific community. So with Christian education. The foundations are laid in the home but the child must move out into the wider family of the Church and so come to understand more fully the pattern of behaviour it has acquired. By being brought into the Church and by being brought up in the worship of the Church the foundations which have been laid in the home can be built upon. In Polanyi's terminology the child will first of all assume that the Church's worship is meaningful and then will by imitating his parents and other Christians indwell the ritual through which comes the vision of God. With regard to actual teaching too we must note that Christian education does not consist in a repetition of a teacher's dictated "right answers". The child and the catechumen must be encouraged to search for and discover that to which the teaching refers. For example with the Bible they must seek to achieve such an heuristic vision of God as it was the Apostles' intention to convey by indwelling their witness.

Yet this indwelling is just like a skill. They have to learn how to understand Holy Scripture, how properly to indwell it by following and imitating the example of those who have already done so. The result of the vision of reality which is focused upon God can only be understood by practising Christians, that is those under control of that vision.[162] Only by learning from those who are already under control of that vision can the formal and informal elements of Christian knowledge be learned.

The Christian faith cannot be formally taught and made completely explicit. Like the knowledge of science it too is a personal knowledge and so the child and the catechumen must tacitly learn by the example of those already in the Church whether parent or ordinary member. The child must dwell in their parents' frame of reference, believe that their parents' pattern of behaviour, worship and prayers are meaningful and seek to find that which gives them meaning. So they will be confronted with the Reality of God to which they must respond in belief or unbelief. Initially through the love and example of their parents the child begins to understand something of the love of God. Their understanding grows wider as it reaches beyond the home into the Church and then steadily builds up to the love of God for all men. In all of this framework in which the child is brought up, the love of God is at first implicit before it is made explicit. It is a tacit knowledge, an informal element that can only be passed on from generation to generation.

This highlights the importance of the Christian upbringing of a child by its parents, and also the danger, ever present, when the informal element is lacking. There must be a continuous tradition where both the formal and the informal elements of faith are passed on from parent to child, from member to catechumen.

Chapter V

ROMAN CATHOLIC AND REFORMED UNDERSTANDING OF AUTHORITY AND TRADITION

1. Historical Origins of Differences

HAVING discussed at some length the role and nature of authority and tradition in the Church and in Science we move on now to compare and contrast the Reformed and the Roman Catholic understanding of authority and tradition and to show the relevance of Michael Polanyi's thought to the debate between them. The main differences between the Reformed and the Roman Catholic understanding are referred to in both "The Dogmatic Constitution of the Church" and the "Decree on Ecumenism" of the Second Vatican Council. In the "Decree on Ecumenism" it is stated that "When Christians separated from us maintain the divine authority of the Sacred Books, they think in a different way from us — and differ among themselves — about the relation of the Scripture with the Church; for in the Church, our catholic faith, asserts that the true magisterium enjoys a unique position when it comes to the exposition and preaching of the written Word of God."[163] This brings out the two main areas where both sides diverge; namely the relation of the Church to Holy Scripture and the understanding of and function of the magisterium in exposition and preaching.

What is at stake here is not the authority of Holy Scripture. What is at stake is the exclusiveness of the authority of Holy Scripture over against all other authorities.[164] The Decree on Ecumenism makes it quite clear, as noted above, that the Roman Catholic Church maintains the divine authority of Holy Scripture. That is not in question. It is the relation of the authority of Holy

Scripture *vis-à-vis* the authority of the Church that was the cause of division at the Reformation and the continued points of debate today. In the light of all we have said so far in this thesis we have seen that the authority of the Church is an authority "under the Word". That is how Karl Barth representing the Reformed understanding, describes the relation between the two. Michael Polanyi's understanding of authority and of tradition in science, is, as this thesis would maintain, of a similar structure to that of the Reformed understanding of authority and tradition in the Church. So to use Michael Polanyi's terminology the church has a "competent" authority but not a "supreme" authority. The Reformed position is one that maintains the exclusiveness of the authority of Holy Scripture over against the authority of the Church. The position of the Reformed Church is summed up in this way by Rudolf Ehrlich. "According to the Reformation understanding, Scripture is self-authenticating and self-interpretative because the Holy Spirit Himself authenticates and interprets its witness. It is for this reason that it is His Word whose unique authority is sovereign and exclusive over against all authorities. It is sovereign and exclusive even over and against the authority of the Church."[165] This assertion made by Rudolf Ehrlich is substantiated by reference to the works of Luther, Calvin and Barth representing the Reformed position.[166]

The Roman Catholic Church, however, understands the relation of the authority of Holy Scripture and the authority of the Church in a quite different way. Whereas the Reformed Church sees Holy Scripture as the norm, in the light of which all else must be tested and judged, the Roman Catholic Church sets up a norm over the norm. It sets its own interpretation of Holy Scripture, contained in its traditions, above Holy Scripture itself.[167] "Is the norm to which the Church of Rome submits Holy Scripture or is the norm the sense and the interpretation which the Roman Church itself has held and does hold?"[168] Since the latter is the case, "This means that the supreme authority to which even the Pope must submit is in actual fact not

Scripture itself but the Roman interpretation of Scripture, the sense of Scripture determined by the magisterium of the Church from which there is no appeal to Scripture itself''.[169]

The Roman Catholic Church would argue that the Apostolic witness was written later in time, that is that the Church existed before it was written down, and that it contains only a limited part of the teaching of the Apostles. There was also an unwritten oral tradition handed down from the Apostles to their successors. Indeed the New Testament itself is only a part of that oral tradition. (Polanyi's understanding of a tacit component in all knowledge is obviously of importance here, and this we will look at in our next section.) On these grounds it is argued therefore that the Bible cannot be the supreme authority. Furthermore while it is maintained that the era of the Apostles was unique and that Revelation ended with them, nonetheless the nature of this distinction is confused by the fact that it is maintained that this unique position is in a sense handed down to the Apostles' successors, to the bishops and direct successors of Peter, whose task was to unfold and clarify the content of that Revelation and allow new truths to come to light. "The Roman Catholic Church distinguishes between a *traditio passiva,* that is a 'treasury of faith' which shall pass down through all ages without changing, and a '*traditio activa*', which constantly unfolds the changeless treasury of faith. It constantly elaborates and develops the content of faith in richer measure, allowing truths not seen before to some to the light of day.''[170]

Thus we have the "*traditio passiva*", the Apostolic witness of the New Testament, and an active tradition which brings new truths to light, and we ascribe to this second source of knowledge the same authority as we do to the first. In this way the Church determines what is and what is not revelation. Alongside the genuine authority of Holy Scripture is placed the authority of the tradition of the Church. But before we go on to look more fully into this distinction between the Reformed and the Roman Catholic understanding of authority and tradition more closely, it might be of value first of all to look very briefly at the historical origins of these differences.

The Council of Trent in 1546 clearly stated the fact that Holy Scripture was one but not the only source of Revelation. It placed alongside Holy Scripture the "tradition" handed down from the Apostles as being a source of knowledge equal in authority to Holy Scripture. What exactly this Apostolic tradition was, was not made clear, and Karl Barth asks "What is this Apostolic tradition which is accepted and has to be heard side by side with Holy Scripture?" The only answer that could be given was that "the universal recognition of the Church shows a definite tradition to be Apostolic, and therefore authentic, and therefore revelation".[171] This is the equivalent of an identification of tradition with Holy Scripture. The tendency towards this identification Karl Barth traces back to the early fathers from Irenaeus to Augustine, although as he points out and as he illustrates with various quotations from these writers, their teaching on the relation of tradition to Holy Scripture was not always consistent and unequivocal.[172] But nonetheless he says "we must note that these statements although they contradict each other, never attain to the clear and critical confrontation of Scripture and tradition which we have in the Reformation decision".[173] Karl Barth would maintain that a certain ambiguity regarding the relation of tradition to Holy Scripture is to be found in the Early Fathers.

With the rise of Gnosticism and Arianism and other Heretical movements in the early Church the question naturally arose in the face of these different interpretations by different groups of individuals, which interpretation was correct? It was in this context that recourse to the "rule of faith" was taken. The rule of faith was a summary of the Christian faith, a summary whose form varied from place to place but whose content was the same. The content of this rule of faith was identical with that of Holy Scripture and was confirmed by Holy Scripture. This rule of faith, this tradition in the Church was taught to every catechumen as a summary of the faith of the Apostles and the faith of the Church. It served as a key to the Holy Scriptures by which they could be rightly understood. Irenaeus for example maintained that scripture had its own design, its own internal pattern and structure and harmony. Heretical interpretations of scripture ignored this pattern in scripture and substituted their own on isolated texts. They rearranged the pattern of Holy Scripture and substituted their own framework of reference. So there was imposed on Holy Scripture a framework of reference which distorted the true meaning of Scripture.[174] It was only through the "rule of faith", through the summary of the faith of the Apostles and of the Church that the true pattern and meaning of Holy Scripture could be discerned. "Scripture could be rightly and fully assessed and understood only in the light and in the context of the living Apostolic Tradition, which was an integral factor of Christian existence. It was so, of course, not because Tradition could add anything to what has been manifested in the Scripture, but because it provided that living context, that comprehensive perspective, in which only the true 'intention' and the total 'design' of the Holy Writ, of Divine Revelation itself, could be detected and grasped. . . . But it was precisely this 'harmony' which could be grasped only by the insight of faith."[175] In other words, the faith of the Church, the rule of faith, the tradition helped to disclose the true intention of Holy Scripture. The Church did not give structure to the Revelation, it was there. It only pointed to that "design" of Holy Scripture.

Thus we have the idea that though Holy Scripture is self-attesting, to counter certain heretical interpretations the "rule of faith", the "scope of faith", the traditions of the Church are invoked. Scripture had to be interpreted in the Church under the guidance of the rule of faith. But what was perhaps, according to Karl Barth, not clearly enough stated in the Early Fathers was the exact relation of tradition and Holy Scripture. There was a certain ambiguity which was the source of future errors.

While there is a need for a living tradition in the Church

where the formal and informal aspects of the Christian faith can be passed on from generation to generation, so that the true intention of Holy Scripture can be discerned, nonetheless this tradition is "under" Holy Scripture. This is the point and the area of ambiguity which Karl Barth takes issue with. Karl Barth argues that by the time of Augustine the inclusion of Holy Scripture itself into the tradition, which was accomplished at a later date, is foreshadowed. "By the turn of the fourth and fifth centuries what the Church as such has to say side by side with Holy Scripture, even if only in amplification and confirmation of it, already has a particular weight of its own, so that the saying which we have already quoted from Augustine — which the Reformers attempted in vain to interpret it *in meliorem partem* — now became possible: in answer to the question what we are to tell those who still do not believe in the gospel, Augustine has to confess, obviously on the basis of his personal experience: "Indeed, I should not have believed the gospel, if the authority of the Catholic Church had not moved me".[176] Whereas Georges Florovsky, I think rightly, interprets this saying of Augustine as not implying the subordination of the Scriptures to tradition, which Karl Barth sees foreshadowed in it, but as rather saying that the gospel is always received in the context of the Church's preaching and teaching, this is not the only possible interpretation nor the one that was indeed placed on it. Augustine's saying could be taken to imply, as Karl Barth would seem to argue, that alongside Holy Scripture and on equal authority with it is the authority of the Church, and not in the sense of the Church's authority being a "competent" authority as we have already discussed.

Despite the ambiguity in this matter among some of the Early Church Fathers from Irenaeus to Augustine, as time goes on the equation of Holy Scripture and tradition becomes clearer. St. Vincent of Lerins, in an attempt to resolve the problem of how different passages might be interpreted differently, put forward as our criterion and norm "We must hold what has been believed everywhere, always and by all", i.e. the criteria of universality antiquity and consent. In other words knowledge of our faith comes through Scripture and tradition. Again St. Vincent's maxim does not, I think, necessarily imply the subordination of scripture to tradition. Indeed we will try to show in our next section how St. Vincent's maxim can be taken as a good description of the way in which the scientific community resolves the problem of knowing which contributions to science are valid and which are not. But again Barth sees foreshadowed in St. Vincent's maxim the subordination of scripture to tradition.

That is what the Council of Trent later made even clearer. The only question that still remained open was the question as to who decided what was universal and had always been believed. The answer implied in St. Vincent according to Barth was that such consent was reached by those who held the teaching office in the Church, the Magisterium. In other words what lay behind the whole development was the "identification of Scripture, Church and Revelation",[177] the roots of which Karl Barth would seem to argue can be traced back to Irenaeus, Tertullian, Augustine and other Early Fathers. There is implied in the identification of the Church and Revelation the idea that the Church stands in a direct and immediate relationship to revelation with the resultant loss of any understanding of the uniqueness of the position of the Apostles. This is what the Reformation totally rejected, this ". . . putting a definite element in Church life which is given the name of divine revelation, the so called tradition, side by side with Holy Scripture, then broadening this element more and more until the whole of Church life seems to be included in it, then subordinating and coordinating Holy Scripture under and with this whole, and finally declaring the whole and therefore itself to be identical with the revelation of God".[178] After the Council of Trent the "Apostolic tradition" was indeed extended to cover all the Church's teaching and historical development. The only question that remained to be clarified was exactly who decided what Scripture is Scripture, what tradition is tradition. In 1870 it was clearly defined that the ultimate infallible and absolute authority of the Church lay with the Papacy. The Pope was the "mouthpiece" of the Church.[179]

The Second Vatican Council to some extent appears to attempt to modify this position and subordinate tradition to scripture. It does not, however, clearly and consistently do this. On the one hand, it is said that "All the Church's preaching, therefore, like the Christian religion itself must be nourished and directed by Holy Scripture",[180] and that "the study of Holy Writ should be, as it were, the soul of theology".[181] Here the Holy Scripture appears to be given a unique place and authority. But this position is compromised when it is said that "In this way it comes about that the Church does not derive from Holy Scripture alone the certainty she possesses on all revealed truths. Therefore both Scripture and Tradition should be accepted with equal sentiments of devotion and reverence. Sacred Tradition and Holy Scripture form a single sacred deposit of the Word of God entrusted to the Church."[182] This leaves open the question as to whether or not tradition contains revealed truth not contained in Holy Scripture.

2. *Scripture* Alone *or Scripture* and *Tradition*

The differences between the Reformed and the Roman Catholic understanding of the relation of scripture and tradition, as outlined above, are often summed up in terms of "scripture *alone*" as against "scripture *and* tradition".

Scriptura sola is the phrase that is taken to sum up the Reformed Church's position. Each individual through searching the Holy Scriptures under the inner guidance of the Holy Spirit is believed to be able to receive the fullness of revealed truth. This is a very individualistic and arbitrary approach to the Christian faith. It is an approach in which the importance of the Church and the authority of the Church's teaching is not given its proper place. This has rightly incurred the criticism of the Roman Catholic Church, and is what Florovsky calls the "sin of the Reformation", exposing Scripture to arbitrary subjective interpretation. He says "Strange to say, we often limit the freedom of the Church as a whole, for the sake of furthering the freedom of individual Christians. In the name of individual freedom, the Catholic, ecumenical freedom of the Church is denied and limited. The liberty of the Church is shackled by an abstract biblical standard for the sake of setting free individual consciousness from the spiritual demands enforced by the experience of the Church."[183] This has often been true in Protestantism.

On the other hand, the position of the Roman Catholic Church is taken to be summed up in the phrase "scripture *and* tradition". In this view there are two sources of revelation. Some of the truths God has revealed are contained in tradition but not in scripture. Scripture and tradition are placed side by side as having equal authority.

The problem would appear to be the emphasis in the Reformed tradition on the distinction between the authority of Holy Scripture and the authority of tradition at the expense of the close connection between them. Whereas in the Roman Catholic tradition the emphasis has been placed on the connection between Holy Scripture and tradition at the expense of the distinction between them. As it has been rightly pointed out "distinction should not be transformed into disconnection, just as connection should not be turned into equation".[184]

Although Holy Scripture has the supreme authority in the Church and is an authority to be distinguished from the authority of the Church's tradition nonetheless they are intimately connected, and this connection must not be lost sight of when an account is given of either. Yet at the same time we must not let the interrelation of Holy Scripture and tradition lead to their being equated. What is needed is a firmer grasp on the nature of the distinction and interrelation. This is an area where Michael Polanyi's insights into the structure of the scientific community sheds some light.

We have noted in our introduction how the scientific community decides what is and what is not science. In reaching this decision many tacit assessments and evaluations are made. Referring to contributions to scientific journals Polanyi says "These principles of judgement that are used to screen articles for publication are largely traditional, since they have been acquired by individual scientists from their mentors and from the literature on their subject, and they are, at bottom, for the most part only tacit understandings. Even when attempts are made to state them explicitly, what these explicit statements *mean* can be known only by scientists in the particular field involved. There is much that cannot be made explicit because it lies at the level of feelings about fitness and in working attitudes that betray an essentially imaginative grasp of how things in that field may be expectd to work or to be."[185] In other words, we must speak of a scientific mind, or attitude, which the layman or student does not yet possess and has to acquire in order to understand that science. There is much that cannot be made explicit, which lies at the level of feelings about the fitness of things and how things ought to be, or a feeling for scientific beauty, originality and plausibility. This "scientific mind" can only be acquired by practice, by imitating an acknowledged expert, by accepting the authority and example of scientists and the scientific tradition.

In a similar way we have to speak of an "ecclesiastical mind", the "mind of the Church", or perhaps better still the "Apostolic mind", a mind which knows what the explicit statements of scripture and of doctrine *mean*. The Apostles have laid the foundations of the Church, an unrepeatable foundation. They and they alone received the Revelation of God in Jesus Christ and witnessed to it.[186] They have laid down the ultimate beliefs which form the premisses of the Church.[187] But this unrepeatable foundation which they have laid has both formal and informal elements. Those who received the Apostles' teaching, in the first instance, received not just explicit statements but also an imaginative grasp of what the Apostles meant by these formal statements. Such a grasp of the Apostles' mind can come only in a similar way to the way in which a student of science has to acquire a "scientific mind", by apprenticing himself to and learning from the authority of a master.

It is in this sense that we can perhaps speak of Apostolic Succession, not in the sense of the Apostolate itself being transmitted from one generation to the next, or in the sense of any new revelation but in the sense of this informal tacit element, the "mind of the Apostles", a tacit understanding of their teaching being learned and passed on from the Apostles to the first converts and from them to the next generation of Christians. A faithful transmission of the Apostolic witness in both its formal and informal aspects. The Reformed Church has perhaps not allowed enough for this "tacit", "living", tradition in the Church in which both the formal and informal aspects of the Christian faith are handed on from one generation to the next. We have noted already that although the Word of God is clear in itself, nonetheless because the Word of God is given *in concreto*[188] in human form in the words of the Apostles, it is capable of being understood or misunderstood by us. In other words, we cannot speak of Holy Scripture "alone" in the rigid sense of the term, we can only speak of Holy Scripture as interpreted according to the "mind of the Apostles" or according to those to whom their message was transmitted and tacitly understood. It may be worth recalling here the fact that, as we mentioned in the introduction, in non-European or American countries science often stagnates or makes little progress because there is no "living" scientific tradition in those countries in which the tacit elements of scientific knowledge can be transmitted and received, in which the scientific mind can be nurtured. It is not enough to have explicit statements alone, to have scripture alone, we need a living tradition in which a proper understanding of scripture is tacitly transmitted and received. This was the reason why, as we noted in our previous section, an appeal was made to the "mind of the Church" or rather the "faith of the Church" in disputes with heretics. Heretics based their arguments on scripture but they rearranged the meaning of scripture, they imposed their own pattern on it. The problem was a problem of interpretation. It was therefore right for an appeal to be made to the "mind of the Church", to tradition, to those who had faithfully received and transmitted the Apostles teaching formally and informally. Tradition did not add anything to Scripture. It provided a "comprehensive perspective" in which the true "intention" of scripture could be grasped[189] and it was only in the Church, the Church built on the foundations laid by the Apostles that scripture could be properly understood. "Tradition was not just a transmission of inherited doctrine, in a 'Judaic manner', but rather the continuous life in the truth. It was not a fixed core or complex of binding propositions, but rather an insight into the meaning and impact of the revelatory events, of the revelation of the 'God who Acts'."[190] This "insight" and "life in the truth" is the tacit element which tradition fosters. The doctrines of the Church are "tools" which must be tacitly used in seeking to understand scripture. There is need therefore for a living tradition in which scripture can be understood.

It is in this sense also that we must understand the saying of Augustine, "I should not have believed the gospel if the authority of the Catholic Church had not moved me". As we have already noted every catechumen must receive their faith from others before them in the Church, from those who already believe and understand. Florovsky makes the point that "He (Augustine) only wanted to emphasise that 'Gospel' is actually received always in the context of Church's catholic preaching and simply cannot be separated from the Church. Only in this context it can be assessed and properly understood."[191] The truth of Holy Scripture is ultimately self-evident but only to those who have received the "insight" which is tacitly received from within the Church. There is no need to interpret this saying of Augustine as Barth does, as implying the subordination of scripture to the traditions of the Church. Again referring to the example of the scientific community there we see that the authority of the scientific community and its traditions are upheld because it is only through indwelling that tradition that the student comes up against the "objective spiritual reality" transcending the tradition. Once tradition has served its purpose of bringing the student into contact with the intrinsic intelligibility of the world he can then judge and perhaps correct the tradition in the light of that reality. The authority of the scientific tradition is not placed on an equal footing with the authority of objective reality itself but it is only through indwelling that tradition that we make contact with the intelligibility of nature and intuit the rationality of the world. We might say that we would not believe in science had the authority of the scientific community and its traditions not convinced us, had we not indwelt their traditions and been brought through them face to face with the truth. This is not to place the authority of the scientific community and its traditions and teachings on the same level as the "objective spiritual reality" transcending the tradition and to which the tradition points, but only to stress that they are indispensable if we wish to understand science. No scientist can work in isolation, alone, cut off from the rest of the scientific community. Even the tools and equipment the present day scientist uses have come from centuries of development from within the scientific community and he has tacitly to learn from that community how to use them skilfully. So too no one in isolation from the Church with Scripture alone could develop far in understanding the Christian faith. The doctrines and traditions of the Church are the tools which he must tacitly use to develop his understanding of scripture. We cannot therefore speak of "scripture *alone*" or of "scripture *and* tradition" (in the sense of both having equal authority). We need to speak of a hierarchy of authority.

We have also noted the fact that every scientific discovery and every new contribution to science have to be assessed tacitly. There are no strict rules which can be impersonally applied to determine what is a valid contribution and what is not, or to determine which scientific theories are correct and which traditions are to be accepted. We noted in our introduction that scientific knowledge is obtained by relying on a set of rules or maxims which are not known as such to the person following them. A skill cannot be accounted for in terms of explicit rules. They serve as a guide only when integrated into the practical knowledge of the art of scientific discovery or of assessment. So it is left to the mind of the scientific community to assess the value and the importance and validity of its traditions and of every new discovery. In the same way within the Church there are no infallible rules, no formal criteria for truth, which can be applied in order to determine what is heresy and what is sound doctrine or a true tradition. This is left to the mind of the Church, a mind steeped in the scriptures. The maxim of St. Vincent which we have looked at is therefore not a formal criterion for testing the truth of the Church doctrine. It is a useful rule and guide, but it does not determine the judgement of the Church in matters of doctrine. The spiritual insight of the mind of the Church cannot be accounted for in terms of this maxim. It is a guide only when it can be tacitly integrated into the practice of this insight. Understood in this sense there can then be made a helpful comparison between St. Vincent's maxim and the way in which the scientific community reaches agreement as to what is a valid contribution to science. St. Vincent enunciates in his famous maxim that we are to hold "that which has been believed everywhere, always and by all" — universality, antiquity, consent. This would seem to be a valid scientific maxim. Every new contribution to scientific knowledge must be universally accepted by a "world wide community of verifiers", by scientists everywhere. There must eventually be a consensus among all scientists, or nearly all, as to its validity. As regards the criterion of antiquity we have noted the fact that "we know more than we can tell", that every discovery reveals an aspect of a reality which will manifest itself in unsuspected ways in the future. It was in this sense that, as we noted in our introduction, Copernicus and Kepler told Newton where to find discoveries which were unthinkable to themselves. A scientific tradition reveals more about a reality than is already known. In other words, there is a permanence in the scientific tradition. It is the *same* reality which each generation is seeking to understand aspects of. Florovsky points out that St. Vincent's maxim is a witness to the "permanence or identity of the *kergyma*, as transmitted from generation to generation".[192] Antiquity was a criterion which was meant to keep out innovations and point to the permanence of the *kergyma*. An authoritative tradition was one which could be traced back to the Apostles and be accepted by the universal consent of the Church. St. Vincent's maxim deals with the origin and preservation of the Christian faith. We must again emphasise, however, that it is not a formal criterion for truth. Universal consent is no guarantee of truth. The scientific community can be mistaken in its judgements. So too can the Church. St. Vincent's maxim is valid only when it is tacitly integrated into the "insight of faith" which can never be made wholly explicit.

3. Function of the Teaching Office

The root cause of such an identification of revelation and tradition appears to be a confusing of revelation itself with the human witness to it, a witness that is meant to point beyond itself to the Reality of the self-attesting, self-authenticating Truth of God itself. This is a confusion of the truth of being with the truth of statements. We are meant to attend *from* our statements *to* the truth, to use Polanyi's terminology, and refine and remake our statements in the light of that to which they refer. Behind this confusion of the truth of being and the truth of statement is a dualistic framework of thought. "The reason for this phenomenon, the corroding or refracting of the referring function of language, in late mediaeval as in modern times, is evidently to be traced to a radical dualism pervading thought, loosening or even detaching its relation to objective reality. . . . Behind that epistemological dualism, of course, lay a thorough going cosmological dualism, in mediaeval times the Ptolemaic Cosmology and in modern times the Newtonian Cosmology."[193] It is just such a dualist framework of thought that Polanyi's philosophy overcomes. He restores our belief in the capacity for human thought to make genuine contact with objective reality and restores our confidence in our own personal judgements as to when we have made such contact. Not to believe in this power of human thought is to open a gulf between our thinking and objective reality. Since this means that we cannot now be persuaded of the truth by the self-attesting authority of the truth itself, as this would be regarded as self-assertion and subjectivity, we must place over ourselves the authority of reason or of experience.[194] Truth then becomes something that is totally impersonal, totally detached, totally objectivist. Out of this dualistic framework of thought comes the positivist concept of science, which Polanyi so strongly rejects, where a scientific theory must not go beyond experience by affirming anything that cannot be tested by experience.[195]

The outcome of the application of a dualistic framework of thought to theology has been the same as that in science. When the self-authenticating nature of the Truth of God has been lost sight of, an attempt is made to overcome the emptiness of self-assertion by again placing over ourselves the authority of the Church or the authority of experience to determine for us what is Truth. The result is confusion of the truth of being with the truth of statement, a confusion of Revelation with tradition, a confusion of the authority of God with the authority of the Church. This confusion is basically the same in the errors of both Neo-Protestantism and Roman Catholicism. It was the Reformed teaching that the Word and the Holy Spirit could not be separated. The Holy Spirit continually confronts the Church with Her Risen Lord in and through the Apostolic witness. The Holy Spirit directs us away from ourselves to the Word. On the one hand, the Neo-Protestant error is that of taking our own subjective opinions and our own self-understanding of the truth and objectivising it and raising it to the level of objective truth. We have here an identification of the Holy Spirit with our own spirituality, an identification of the Holy Spirit with subjectivity. We are confronted not with the Truth but with our own interpretations of that truth. The error is a similar one in Roman Catholicism. Here the collective subjectivity of the Church, the Church's understanding of its faith is objectivised and exalted to the level of objective truth. The Holy Spirit and the Word are again separated and the Holy Spirit identified with subjectivity. On the one hand, each individual gives expression to his faith and such religious experience is equated with the Truth and on the other hand the mind of the Church, the teaching of the Magisterium replaces the Holy Spirit and is equated with the Truth. In both cases the Church is no longer confronted with the Word of God, with self-authenticating, self-attesting Truth. Its own understanding of the Truth, or the individual's understanding of the truth is objectivised and put above the level of objective truth. It was against such an error that the Reformers stressed their doctrine of "justification by grace". "When we apply justification by Grace to the task of theology it means that we can never claim the truth for our own statements, but must rather think of them as pointing away to Christ who alone is the Truth. Theological statements do not carry their truth in themselves but are true only in so far as they direct us away from ourselves to the one Truth of God. That is why justification is such a powerful statement of objectivity in theology, for it throws us at every point upon God himself, and will never let us repose upon our own efforts."[196]

All of this has a very important bearing on our understanding of the debate between the Reformed Church and the Roman Catholic Church over their understanding of authority. It was for example Kung's criticism of Luther that Luther took to himself an authority belonging to the Church. "The actual reason for (Roman Catholics) rejecting Luther was this: For all that it included genuine reforms, and despite his conservatism, often stressed today, Luther's Reformation was essentially a Revolution. He brought the very essence of the (Roman) Catholic Church into question when (this was the real innovation) he set his personal, subjective and yet (by his intention) universally binding interpretation of the Scripture in principle above the Church and her tradition."[197] The Roman Catholic criticism of the Reformation was that it led to arbitrariness, to individuals giving their own subjective interpretation of Scripture and then claiming for these a universal acceptance. To the Roman Catholic Church the subjectivity of individuals had become the rule of faith for everyone else, and such an approach lacked any established principle of authority which could say universally how Holy Scripture should be interpreted and understood. "For the Roman Catholics the authority of the Reformation Church is merely the authority of Luther or of Calvin or of any other preacher whose subjective interpretation of Scripture is put in principle above the Church and its tradition."[198] But this accusation of making the subjective interpretations of an individual a binding standard or criterion of truth was totally rejected by the Reformers. They claimed instead to have discovered the true authority of the Church. "Since the Spirit sends the Church continually to the Word as the witness of the Prophets and Apostles and illumines it so that the Church hears the voice of Christ Himself, the Church has authority under the Word to proclaim the Gospel of Jesus Christ, being able in so doing to distinguish between the Word of God and the word of man."[199] The Reformed position was one that did not separate the Word and the Holy Spirit, and therefore since the Church was directed away from itself to the Truth, the Reformers were not presenting a subjective interpretation of Scripture. Luther following Augustine, to whom it might be noted Michael Polanyi often refers, speaks of the soul being apprehended or overwhelmed by the truth, so that we are not able to judge the truth but only to say yes that this is the truth. The Holy Spirit illumines, enlightens the Church by pointing it to the Word and so, being confronted by the Truth, the Church can only say "yes" to it. Luther in the Babylonian Captivity of the Church says "But, as Augustine says elsewhere, the truth itself lays hold on the soul and thus renders it able to judge most certainly of all things; however the soul is not able to judge the truth, but is compelled to say with unerring certainty this is the truth. For example our mind declares with unerring certainty that three and seven are ten; and yet it cannot give a reason why this is true; although it cannot deny that it is true. It is clearly taken captive by the truth; and, rather than judging the truth, it is itself judged by it."[200] It is in this way that the Reformers rejected any idea that theirs was merely a subjective interpretation of Holy Scripture and at the same time it was this argument that formed the basis of their criticism of the Roman Catholic position. The Roman Catholic Church was accused by the Reformers of having identified the Holy Spirit with their own spirit, with the result that the Roman Catholic Church was confronted not with Holy Scripture, with the Truth, to which the Holy Spirit was meant to direct us, but they were confronted instead with their own interpretation of Holy Scripture. There was in this way given to the Church an absolute authority that belonged only to Holy Scripture, an authority which the Roman Catholic Church defended by arguing that the promised guidance of the Holy Spirit guaranteed the infallibility of the Church, the infallibility of the mind of the Church, as expressed by the Magisterium. To both this collective and individual subjectivity the Reformers objected and put in its place the unbreakable bond between the Word and Holy Spirit, directing us away from ourselves to the Truth. It was in this way that the Reformers overcame the objective/subjective dichotomy and the dualistic framework behind it.

Michael Polanyi's philosophy is again of relevance to the debate between the Reformed and Roman Catholic Church in their different understanding of authority and tradition. The Roman Catholic criticism of the Reformers' position as being one which elevates to a universal status the subjective views of individuals, based on no principle of authority for such views, is particularly illumined by Michael Polanyi's thought. We can say that the Reformers had a personal faith, a personal knowledge in the same sense in which Polanyi speaks of Personal Knowledge. Polanyi's concept of Personal Knowledge transcends as we have seen the subjectivist/objectivist approach to science and restores the place of the knowing person in knowledge. The scientist's knowledge and understanding of the world is not subjective. It is his personal assent to what is given, to a "spiritual reality" that stands over him and compels his assent. It is the act by which he himself indwells the subsidiary clues and integrates them to the focal. And although the scientist chooses his own problems and pursues them in his own way he is not a law unto himself. Because he believes in the intelligible order inherent in the world and in man's capacity to know and understand that order he submits himself to the self-attesting authority of that order. His own thinking must be shaped and moulded and called in question by what is there. There is a "spiritual reality" that stands over him and compels his allegiance and responsible commitment. This we have already fully dealt with. But there is a parallel here between Polanyi's understanding of scientific belief as a personal knowledge and the Reformed understanding of an individual's understanding of Holy Scripture. The interpretation of Holy Scripture that Luther or Calvin taught is not subjective as the Roman Catholic Church argued. It is personal. It is the assent given by each individual to the self-attesting Truth of God, as people are directed away from themselves to that Reality in the Holy Spirit. There is no question of any subjective individualism on the part of the Reformers, just as there is no question of a subjective individualism on the part of a scientist. Indeed the Reformers had to persuade and convince the rest of the Church of the rightness of their interpretation. They did so not by any criteria for judging the Truth but by pointing others to that Truth, that others confronted by it might be persuaded. They assumed each individual's capacity to know the Truth of God, by the Grace of God. John Knox in the Scots Confession writes ". . . if any man will note in this our confession any article or sentence repugnant to God's Holy Word that it would please him of his gentleness and for Christian charity's sake to admonish us of the same in writing; and we upon our honour and fidelity, by God's grace do promise unto him satisfaction from the mouth of God, that is, from His Holy Scriptures, or else reformation of that which he shall prove to be amiss."[201] There is here no subjective individualism but rather a framework similar to what Michael Polanyi describes as "overlapping neighbourhoods" in the scientific community where over and above each scientist is a transcendent reality to which each scientist refers and in the light of which each scientist judges and assesses each other's work. So too the Reformers not only had to persuade and convince the rest of the Church they had to listen to and accept the criticism of their teaching made by other Christians. This they did because there was an actual Word of God to the Church, over and above the Church, to which they pointed. There is a "spiritual reality" transcending the scientific tradition and a Word of God transcending the Church's tradition. This tradition must not be confused with the reality to which it points. Such was the error of Roman Catholicism.

Furthermore we must add that no magisterium, or single person, or single group or any "specific" authority, can determine or dictate what is the truth. Although Polanyi speaks of "personal" knowledge he does so only from within the community of scientists. No scientist, despite his personal creativity and insight can work outside the scientific community. He dwells in the scientific tradition handed on to him, he works by the principle of "overlapping neighbourhoods" and accepts and respects the authority of the scientific community and upholds it. Authority resides not in an individual or group but in the whole community of scientists.

So too in the Church. It is the task of the whole Church to determine sound doctrine and tradition. It is not the task of an individual or of a group or magisterium. Because we believe in the capacity of each individual within the Church to know the Truth of Christ the authority resides in the consensus of the whole Church. It is true that some within the Church as in the scientific community may possess a greater degree of authority than others, but this is an authority given to them by the scientific community or Church, to speak on behalf of and in the name of that community. As Florovsky says "He (the Bishop) must speak not from himself, but in the name of the Church".[202]

Therefore authority belongs to the whole Church. It is a "general" authority established through debate and always engaged in "under the Word".

CHAPTER VI

THE DEVELOPMENT OF TRADITION

WE move on now to another area of disagreement between the Reformed and the Roman Catholic Church, namely that as a result of the Roman Catholic Church placing tradition alongside Holy Scripture as an equal authority it has formulated dogmas and creeds not contained in the witness of Holy Scripture, belief in which are deemed necessary for salvation. We come to the question of the development of the tradition. At the outset in must be noted that the Reformed Church did not reject the idea of doctrinal development, the question that is at stake here is how we are to understand the development of a tradition.

Rudolf Ehrlich points out that the two-source theory of Revelation, that revelation is contained partly in Holy Scripture and partly in tradition, the view which was held by Post Tridentine Roman Catholic Theologians, allowed the Roman Catholic Church to explain and justify dogmas not in Holy Scripture. The Immaculate Conception and the Bodily Assumption of Mary would be two such dogmas. They can be explained by Post-Tridentine theologians by pointing to the oral tradition[203] and by claiming that the Church is guided by the Holy Spirit in all its doctrinal pronouncements. In this case the Reformed Church which holds Holy Scripture alone to be the sole witness to Revelation must reject outright such dogma. It does not accept tradition as of equal authority to Holy Scripture. But, as Rudolf Ehrlich further points out, difficulty arises when in the ecumenical dialogue between the Reformed and Roman Catholic Church certain Roman Catholic theologians do not hold the two-source theory of revelation but seek to maintain instead that all Church dogma and doctrinal formulations come from one source, i.e. Holy Scripture.[204] This is the view of Geiselmann who sees Holy Scripture sufficient as regards to its content but who would argue that it needs to be interpreted and that this is the function of the living tradition of the Church. This is the understanding that has been developed by the Tübingen school and the one we seek to question. It would also appear to be the teaching of the Second Vatican Council. Referring to the magisterium it is said that "this magisterium is not superior to the Word of God, but ministers to the same word by teaching only what has been handed on to it, in so far as, by divine command and with the assistance of the Holy Spirit, the magisterium devoutly hears, religiously keeps and faithfully explains the word, and from this one deposit of faith derives all those things which, it proposes to us for acceptance as divinely revealed".[205] How can those in the Roman Catholic Church, who take Holy Scripture as the sole source of revelation, and who see the task of the tradition of the Church as being one of interpreting that revelation, possibly justify the dogma of the Immaculate Conception and the Bodily Assumption of Mary? Rudolf Ehrlich describes one attempt to account for this doctrinal development by referring to the argument put forward by Karl Rahner.[206] It is Rahner's contention that all new doctrinal formulations are already contained in an earlier form of the Church's faith. Rahner defines what he means by "contained" in the following way. He argues that the later creeds and dogmas are developed from what is implicit in earlier knowledge by maintaining that this earlier knowledge does not always consist only of explicit statements. There are certain "experiences" for example about which we "know" more than we can state.[207] How similar Rahner's terminology appears to be here to that of Michael Polanyi when he maintains that in skills we know more than we can tell and can only make known focally later on what we were subsidiarily aware of earlier. In this Rahner maintains that later knowledge does not develop logically from earlier knowledge or from earlier statements — but are "the formulation for the first time of propositions about an already possessed knowledge . . . Christ as the living link between God and the world, whom they (the Apostles) have seen with their eyes and touched with their hands, is the objective content of an experience which is more elemental and concentrated, simpler and yet richer than the individual propositions coined in an attempt to express this experience — an attempt which in principle can never be successful . . .".[208] In this way what is handed on in the Apostolic Succession is a "living experience", the Holy Spirit present in the Church. The dogma and doctrine developed by the Church may not formally be deduced from the earlier statements of the New Testament but they are "compresent", "communicated" in the witness of the Apostles. What was implicit in their experience is only now being made explicit, and under the guidance of the promised Holy Spirit the Church is prevented from making any false doctrinal statement. Thus it is why, Rahner can maintain that dogma of the Immaculate Conception and Bodily Ascension of Mary have a scriptural source. They were "compresent" in the Apostolic experience which has been passed on to their successors. This whole explanation of doctrinal development involves the prolongation of the Apostolic experience in the Church, a concept which really only covers over the fact that the Catholic Church is inserting new elements into the faith springing from its own subjective understanding.

An alternative way of looking at the development of tradition which has also been put forward in the Roman Catholic Church is the view that maintains that the tradition grows like an organism; it stays the same but grows and develops. This is the view put forward by Karl Adam, "We Catholics acknowledge readily, without any shame, nay with pride, that Catholicism cannot be identified simply and wholly with primitive Christianity, nor even with the gospel of Christ, in the same way that the great oak cannot be identified with the tiny acorn. There is no mechanical identity, but an organic identity. And we go further and say that thousands of years hence Catholicism will probably be even richer, more luxuriant, more manifold in dogma, morals, law and worship, than the Catholicism of the present day."[209] Here we see more clearly the importance of the distinction discussed earlier between a *traditio passiva* and a *traditio activa*. The *traditio activa* constantly develops the *traditio passiva*. The tradition is like a seed growing and developing into a tree. It is a living thing which even though it remains organically the same yet must grow and develop. In accordance with this view the Apostles become one link in the tradition. They have no unique position, apart from being the first link. The Apostolic witness is not nomative in the sense in which we have already discussed.

Both of these alternative views of the developing tradition in the Roman Catholic Church are rejected by the Reformed Churches. But though it rejects both of these attempts to explain the development of the tradition it does not reject the concept of development of doctrine and dogma itself. It does not have a view of tradition which sees it as a static entity. It is the understanding of the nature of this developing tradition that is what is at stake. The doctrine of the Trinity expresses what is implicit in the New Testament witness and does so in a form and method of expression relevant to the particular problems the Church faced in the third and fourth centuries. Each generation, if it can better express the truth of the Biblical witness must develop the dogma and doctrine of the Church. It must use the language and terminology of its day to express the truth of the Apostolic witness in new formulations which are better and more adequate than the older ones. Karl Barth describes this understanding of the development of the tradition in this way; he describes how in certain circumstances and certain situations the Church is confronted with "different expositions of one and the same theme."[210] This then means that the "existing confession of the common faith and therefore the existing exposition and application of Holy Scripture is called in question because the unity of the faith is differently conceived and there is such different teaching on the basis of the existing unity that the unity is obscured and has to be rediscovered. The expression which is valid and which was once really the expression of unity no longer suffices."[211] Thus there arises a new confession, new doctrine and dogma. This new confession follows the direction of the old confession and better expresses for us the truth to which the old confession pointed us in the new situation in the Church. Indeed, as we have already mentioned, as we follow the direction of the old confession it itself may demand its own correction and development in the light of that to which it points. The Church must always in this way be seeking to improve its doctrine and dogma.

The main point to notice, however, in the Reformed understanding of doctrinal development is that the Church's doctrines must be implicit in the Apostolic witness. The material content of the doctrine must be contained in Holy Scripture. Such is the case with the doctrine of the Trinity but not with the doctrines of the Immaculate Conception or the Bodily Assumption of Mary. "What is not found in Scripture and had therefore to be 'developed' are the precise dogmatic statements on the Trinity which formalise (as far as that is possible) what is materially contained in the witness of the Prophets and Apostles. These dogmatic statements were (and are being) made by the Church as a result not of a development of the truth but of a development in the understanding of the truth."[212] It is in this sense that the Church must always be making and remaking, developing its traditions, its dogmas and creeds. It has to develop its understanding of the Truth and speak to its contemporary situation.

The disagreement between the Reformed Church and the Roman Catholic Church is then not over the development of tradition but over what is developed — "the understanding of the Truth or the truth itself". The Reformed position clearly states that all doctrine and doctrinal development are "under" Holy Scripture; and the clarification of this tradition as new situations face the Church are also "under" Holy Scripture. It would argue that the Roman Catholic Church understands that development of tradition in an erroneous way. There is no Biblical evidence for the doctrines of the Immaculate Conception or the Assumption of Mary. This is a development of the Truth, a development that is regarded as being infallible by the Catholic Church because of the promised guidance of the Holy Spirit, and explained either by regarding tradition as a growing organism or by Rahner's theory these doctrines being "com-present" and "communicated". Thus even though they are not in Holy Scripture they are accepted by the Catholic Church. The Reformed criticism of this would be that the Roman Catholic Church has identified its own spirit with the Holy Spirit, so that it is confronted not with Holy Scripture but its own subjective self-understanding of Holy Scripture.

Again it is highly illuminating to look at how Polyani describes the development of tradition in science. Here we see a parallel between the Reformed understanding of development of tradition and Polyani's understanding of the same matter in science. As we have discussed previously there are no strict rules explicitly laid down by which we can discover and demonstrate the truth of scientific propositions. Instead the scientific community has to judge the value of these scientific propositions by certain self-set standards, standards which can only be transmitted from one generation to the next by learning the skill in which these standards are embodied. This leaves room for reinterpretation of these standards. "It follows that every process of reinterpretation introduces elements which are wholly novel; and hence also that a traditional process of creative thought cannot be carried on without wholly new additions being made to existing tradition at every stage of transmission."[213] Thus a tradition of thought like science cannot but be creative. The tradition of science is one "that upholds an authority that cultivates originality".[214] This is a consequence of the belief that each scientist has the ability to recognise an objective reality, the transcendent rationality of the natural world. Their scientific propositions only intimate, only point beyond themselves to that reality, revealing aspects of it. Hence the significance of that phrase which recurs again and again in Polanyi's writings "We know more than we can tell". Hence also why for example both Copernicus and Kepler were able to tell Newton where to make discoveries unthinkable to themselves. "Copernicus anticipated in part the discoveries of Kepler and Newton, because the rationality of his system was an intimation of a reality incompletely revealed to his eyes. Similarly, John Dalton (and long before him the numerous precursors of his atomic theory) beheld and described the dim outline of a reality which modern atomic physics has since disclosed in precisely discernible particulars."[215] From this follows Polanyi's definition of reality as that which is capable of manifesting itself in unsuspected ways in the future and how he sees existing scientific knowledge as serving as subsidiary clues to future discoveries. In other words, what we have in science is not a development of truth but a development in the scientist's understanding of the truth, a development in his understanding of the objective reality before him. For example quantum mechanics does not contradict classical mechanics (although it does correct it). These are not two different truths either referring to two different realities or contradicting each other. Rather is it the case that in the newer quantum mechanics we regain the old concepts of classical mechanics as a limiting case when velocities are small.[216] There has been a development in the scientists' understanding of the truth but not in the truth itself. This is very similar to our previous discussion when we looked at how a new confession arose. The new confession followed the direction of the old confession and better expressed for us the truth to which the old confession pointed. There is no contradiction here nor any new truth being added.

The ultimate beliefs in science therefore remain the same, but the natural laws which are open to the ordered pattern in nature, which the scientist believes in, can develop, and bring forth new aspects of that ordered pattern in nature. As a consequence of this the scientist cannot accept what does not derive from his enquiry into objective reality, into the truth. The truth itself, nature itself, must call in question his own ideas about it, his own presuppositions about it. Indeed the experiments which the scientist performs are really questions which he puts to nature — questions which nature herself must answer yes or no to.

Science is truly objective in the sense that we are dealing with an intelligibility which we cannot control or manipulate. It transcends all our thinking about it. It continues to reveal itself to us in unsuspected and unanticipated ways. We are up against a transcendent intelligibility which "commands our respect", "arouses our admiration" to which our minds have to submit and under whose impact form concepts that reflect that rationality.

We can apply Polanyi's definition of reality to theology. Theology is truly objective in the sense that we are dealing with the ultimate Reality of God, whom we cannot control or manipulate or classify. In Christ, we are confronted with Someone who transcends all our thinking and speaking about Him. The Biblical statements about him point to truths "greater than they can tell" and indicate more than can be formally expressed. We noted earlier that the Incarnation and Resurrection form two of our ultimate beliefs. These are beliefs which transcend anything we can say about them, as they display a depth of intelligibility which outruns our grasp of it. They point to "more than they can tell" and indicate far more than can be expressed. As in science, where we are dealing with a reality over which we have control and which can continue to reveal new aspects of itself in unsuspected ways in the future (but always in coherence with what has been disclosed before) so too in theology. In theology we are dealing with the Reality of God, with "an objectivity with an infinite power of self-revelation beyond all our expectations in the future".[217] There is then in theology a deepening tradition in which new features and aspects of that once for all revelation are being brought to light. Though we must stress that this tradition is one which is grounded and dependent on the Apostolic witness and must always be in coherence with it, with what has been disclosed there. Florovsky points out that not everything in the Church dates from Apostolic times. But then he goes on to say that this does not mean something has been revealed which was unknown to the Apostles, "Revelation has not been widened, and even knowledge has not increased. The Church knows Christ now no more than it knew Him at the time of the Apostles. But it testifies to greater things. In its definitions it always unchangeably describes the same thing, but in the unchanged image ever new features become visible. But it knows the truth not less and not otherwise than it knew it in time of old."[218]

A genuine Church tradition will always be one which is creative and goes on disclosing new features of the Revelation of God in Jesus Christ. But again we must stress that, as in the scientific tradition where nature herself must answer yes or no to the experimental questions that scientists put to her, so too in theology the Apostolic witness must answer yes or no to what is contained in the Church's tradition. There must be coherence with what has been disclosed in the Apostolic witness.

We can perhaps illustrate this with the doctrine of the Trinity. This doctrine is not some new truth or new revelation which has been given to the Church. What we have here is a development in the Church's understanding of the Truth to which the Apostolic witness points, a witness which indicates far more than can be expressed at the time. It is a development grounded in the Apostolic witness and in agreement with what is disclosed there.

SUMMARY

IN Jesus Christ we are confronted with the reality of God's self-revelation. This is our conviction which we must plainly state at the outset. We cannot in a detached impersonal way demonstrate the truth of this conviction. Jesus Christ carries with Him His own authority which calls forth from us our convictions and belief. We are submitting ourselves to the self-authenticating authority of the Truth itself.

We are confronted with the ultimate authority of God's self-revelation only in and through the Biblical witness of the Prophets and Apostles. Here once again Polanyi's philosophy was of relevance. In the Apostolic witness we are dealing with two levels of reality. The first of these levels is the Incarnation and Revelation of God in Jesus Christ. This level organises and gives distinctive shape, form and meaning to the lower level, that of the Apostolic witness; a witness which can only be understood as it is correlated with that higher level. This understanding comes as we indwell the Biblical witness and attend from it to the vision of God which it was the Apostles' intention to convey. It is here that due regard must be paid to the tacit, informal element in the Apostolic witness. A student of science must learn what scientific statements mean by watching and imitating a master of science in the way in which he practises his skill, chooses problems, reacts to new clues and unforeseen difficulties. Only in this way is a vision of what scientific statements mean transmitted to him. He must acquire in this way a scientific "mind". Without this living tradition in science in which this tacit element can be passed on from master to pupil science would stagnate. The same is true of the Church. What we require in the Church is not simply the explicit statements of the Apostles but also a tacit understanding of their teaching being passed on from them to their immediate successors and from them to succeeding generations. The Apostles therefore initiate a "living" tradition within which their witness is understood. They lay the foundations of the Church, the community of those who accept their understanding of the meaning and significance of the life of Christ. The Church has a "competent" authority in that it understands the "mind" of the Apostles, an authority which must be respected by the catechumen and by individual members of the Church themselves. We are dealing here with a hierarchy of authority. The teaching, traditions and doctrines of the Church must be regarded as "tools" to be tacitly used to help us understand the intention of Holy Scripture. They serve their purpose when they confront us with and point us to the revelation of God in Christ to which the Apostles witness and which is our primary authority, and have to be judged, and if need be corrected, in the light of that to which they refer us. Furthermore since we are dealing with the Revelation of God, with knowledge of God, we are dealing with a reality that far outruns our understanding. According to Polanyi's definition of reality as being that which is able to manifest itself in surprising and unsuspected ways in the future so too with the Reality to which theology refers. The Apostolic witness points to more than it can tell. This allows for a deepening and developing tradition in theology where new aspects of the once and for all revelation in Christ are brought to light but always in coherence with what was disclosed in the Apostolic witness.

The nature and role of authority and tradition in science are of a similar structure to the nature and role of authority and tradition in theology. Indeed both exist and continue to exist because both can embody their premisses in a tradition which is held in common by a community which has the authority to uphold and interpret, apply and confirm them. The "spiritual reality" transcending and embodied in this tradition is the ultimate authority which calls forth our responsible commitment to it.

BIBLIOGRAPHY

Barth, K., *Church Dogmatics,* 1:2 (Edinburgh, T. and T. Clark, 1956).

Chadwick, H., Yarnold, E. J., *Truth and Authority, A Commentary on the Agreed Statement of the Anglican-Roman Catholic International Commission* (S.P.C.K., 1977).

Church of Scotland Panel on Doctrine, *Agreement and Disagreement* (Edinburgh, Scottish Academic Press, 1977).

Duncan, L. H., *Calling All Parents* (Edinburgh, St. Andrews Press, 1978).

Ehrlich, R. J., *Rome — Opponent or Partner* (London, Lutterworth Press, 1965).

Florovsky, G., *Bible, Church, Tradition: An Eastern Orthodox View, Collected Works,* Vol. 1 (Belmont, Nordland Publishing Company, 1972).

Gelwick, R., *The Way of Discovery: An Introduction to the Thought of Michael Polanyi* (New York, Oxford University Press, 1977).

Jaki, S. L., *The Relevance of Physics* (Chicago, University of Chicago Press, 1966).

Jaki, S. L., *The Road of Science and the Ways to God* (Edinburgh, Scottish Academic Press, 1978).

Kuhn, T. S., *The Structure of Scientific Revolutions* (Chicago, University of Chicago Press, 1962).

Langford, T., "Michael Polanyi and the Task of Theology", *Journal of Religion,* 46 (January 1966) pp. 45-55.

Osborn, R. T., "Christian Faith as Personal Knowledge", *Scottish Journal of Theology,* Vol. 28, No. 2.

Polanyi, M., *Knowing and Being* (London, Routledge and Kegan Paul, 1969).

Polanyi, M., *The Logic of Liberty* (London, Routlege and Kegan Paul, 1951).

Polanyi, M. and Prosch, H., *Meaning* (Chicago and London, The University of Chicago Press, 1975).

Polanyi, M., *Personal Knowledge* (London, Routledge and Kegan Paul, 1973 edit.).

Polanyi, M., *Science, Faith and Society* (Chicago and London, The University of Chicago Press, 1973 edit.).

Polanyi, M., *The Tacit Dimension* (London, Routledge and Kegan Paul, 1967).

Polanyi, M., *Scientific Thought and Social Reality,* New York, *Psychological Issues,* Vol. VIII, Number 4, Monograph 32.

Polanyi, M., *The Study of Man* (Chicago and London, The University of Chicago Press, 1959).

Polanyi, M., "Faith and Reason", *The Journal of Religion,* 1961, Vol. XLI, Number 4.

Schilpp, P. A. (Edit.), *Albert Einstein: Philosopher Scientist* (Evanston, Library of Living Philosophers, 1949).

Second Vatican Council, *Dogmatic Constitution of the Church* (London, Catholic Truth Society, 1966).

Second Vatican Council, *Dogmatic Constitution on Divine Revelation* (London, Catholic Truth Society, 1966).

Second Vatican Council, *Decree on Ecumenism* (London, Catholic Truth Society, 1966).

Skydsgaard, K. E., *One in Christ* (Muhlenberg Press, 1957).

Torrance, T. F., "Truth and Authority", *The Irish Theological Quarterly Review,* Vol. XXXIX, No. 3, July 1972.

Torrance, T. F., "The Place of Michael Polanyi in the Philosophy of Modern Science", *Ethics in Science and Medicine,* Vol. 7, No. 1, 1980.

Torrance, T. F., *Theology in Reconciliation* (London, S.C.M. Press Ltd., 1965).

Torrance, T. F., *Space, Time and Resurrection* (Edinburgh, The Handsel Press, 1976).

Torrance, T. F., *God and Rationality* (London, Oxford University Press, 1971).

Torrance, T. F., *Royal Priesthood, Scottish Journal of Theology, Occasional Papers,* No. 3 (Edinburgh, Oliver and Boyd, 1963).

REFERENCES

1. Michael Polanyi (with H. Prosch), *Meaning*, Chicago, 1975, pp. 3–4.
2. Michael Polanyi, *Personal Knowledge*, London, 1958, pp. 6–8.
3. *Ibid.*, p. 7: p. 134.
4. Michael Polanyi, *Scientific Thought and Social Reality*, New York, 1974 and Oxford, 1977, p. 113.
5. Richard Gelwick, *The Way of Discovery*, New York, p. 50.
6. F. S. C. Northrop, "Einstein's Conception of Science", *Albert Einstein: Philosopher Scientist* (edit by P. A. Schilpp), New York, 1949, p. 387.
7. *Ibid.*, p. 388.
8. *Ibid.*, p. 394.
9. T. F. Torrance, "The Place of Michael Polanyi in the Modern Philosophy of Science". *Ethics in Science & Medicine*, Oxford, 1980, Vol. 7, p. 60. (Reprinted in *Transformation and Convergence in the Frame of Knowledge*, Belfast and Grand Rapids, 1984, pp. 107–173.)
10. Michael Polanyi, *Scientific Thought and Social Reality*, p. 80. *Science, Faith and Society*, Chicago, 1964, p. 89.
11. Michael Polanyi, *Scientific Thought and Social Reality*, p. 80. *Science, Faith and Society*, p. 9.
12. Stanley L. Jaki, *The Road of Science and the Ways to God*, Chicago & Edinburgh, 1978, p. 87.
13. *Ibid.*, p. 87.
14. Michael Polanyi, *Meaning*, p. 31.
15. *Ibid.*, p. 31.
16. Michael Polanyi, *Knowing and Being*, London, 1969, pp. 81ff. *Science, Faith and Society*, p. 49.
17. Michael Polanyi, *Scientific Thought and Social Reality*, p. 71.
18. Richard Gelwick, *The Way of Discovery*, p. 14.
19. Michael Polanyi, *Scientific Thought and Social Reality*, p. 101.
20. *Ibid.*, p. 113.
21. Michael Polanyi, *Personal Knowledge*, p. 9.
22. Michael Polanyi, *Scientific Thought and Social Reality*, p. 101.
23. Michael Polanyi, *Meaning*, p. 31.
24. Michael Polanyi, *Knowing and Being*, pp. 114ff.
25. *Ibid.*, p. 128.
26. *Ibid.*, p. 138.
27. Michael Polanyi, *Science, Faith and Society*, p. 33.
28. Michael Polanyi, *Knowing and Being*, p. 140.
29. Michael Polanyi, *Meaning*, p. 38.

30. Michael Polanyi, *Personal Knowledge*, p. 96.
31. Michael Polanyi, *Knowing and Being*, p. 110.
32. Michael Polanyi, *The Study of Man*, London, 1959, p. 37.
33. Michael Polanyi, *Personal Knowledge*, p. 256.
34. *Ibid.*, p. 300.
35. *Ibid.*, p. 305.
36. Michael Polanyi, *Science, Faith and Society*, p. 41.
37. Michael Polanyi, *Knowing and Being*, p. 148.
38. See p. 13.
39. Michael Polanyi, *The Tacit Dimension*, London, 1967, p. 15.
40. Michael Polanyi, *The Study of Man*, p. 74.
41. Michael Polanyi, *Knowing and Being*, p. 152.
42. Michael Polanyi, *The Tacit Dimension*, p. 33.
43. Michael Polanyi, *Knowing and Being*, p. 218.
44. Michael Polanyi, *The Tacit Dimension*, p. 41.
45. Michael Polanyi, *Personal Knowledge*, p. 374.
46. *Ibid.*, p. 375.
47. *Ibid.*, p. 376.
48. Michael Polanyi, *The Study of Man*, p. 99.
49. Michael Polanyi, *Personal Knowledge*, p. 378.
50. Michael Polanyi, *Knowing and Being*, p. 94.
51. Michael Polanyi, *Personal Knowledge*, p. 204.
52. This is dealt with in *Science, Faith and Society* and in the first three essays of *Knowing and Being*.
53. Michael Polanyi, *Meaning*, p. 184.
54. Michael Polanyi, *Personal Knowledge*, p. 50.
55. Michael Polanyi, *Science, Faith and Society*, p. 15.
56. Michael Polanyi, *Personal Knowledge*, p. 53.
57. Michael Polanyi, *Science, Faith and Society*, p. 44.
58. *Ibid.*, p. 43.
59. Michael Polanyi, *The Tacit Dimension*, p. 60.
60. Michael Polanyi, *Science, Faith and Society*, pp. 43f.
61. *Ibid.*, p. 43.
62. Michael Polanyi, *Personal Knowledge*, p. 165.
63. Thomas S. Kuhn, *The Structure of Scientific Revolutions*, chap. 5.
64. Michael Polanyi, *Science, Faith and Society*, p. 46.
65. Michael Polanyi, *Personal Knowledge*, pp. 287f.
66. *Ibid.*, p. 294.
67. *Ibid.*, p. 156.
68. Michael Polanyi, *Knowing and Being*, p. 50.
69. *Ibid.*, p. 52.
70. *Ibid.*, p. 50.
71. Michael Polanyi, *Science, Faith and Society*, pp. 47ff.
72. Michael Polanyi, *Knowing and Being*, pp. 53ff.
73. *Ibid.*, pp. 74ff.
74. Michael Polanyi, *Knowing and Being*, pp. 81.
75. *Ibid.*, pp. 97ff.

76. *Ibid.*, p. 54.
77. *Ibid.*, p. 67.
78. *Ibid.*, pp. 83ff.
79. *Ibid.*, p. 65.
80. *Ibid.*, p. 66.
81. Michael Polanyi, *Personal Knowledge*, p. 160.
82. Michael Polanyi, *Science, Faith and Society*, pp. 51–52.
83. Michael Polanyi, *Knowing and Being*, p. 70.
84. Michael Polanyi, *Science, Faith and Society*, p. 54.
85. Richard Gelwick, *The Way of Discovery*, p. 117.
86. Michael Polanyi, *The Tacit Dimension*, p. 61.
87. *Ibid.*, p. 62.
88. Michael Polanyi, *Personal Knowledge*, p. 299.
89. *Ibid.*, p. 269.
90. *Ibid.*, p. 267.
91. *Ibid.*, p. 305.
92. *Ibid.*, p. 311.
93. Thomas F. Torrance, *Space, Time and Resurrection*, p. 19.
94. Michael Polanyi, *The Tacit Dimension*, p. 61.
95. Thomas F. Torrance, "Truth and Authority", *The Truth, Theological Quarterly*, XXXIX, 3, 1972, p. 215. (Reprinted in *Transformation and Convergence in the Frame of Knowledge*, Belfast and Grand Rapids, 1984, pp. 303–332.)
96. *Ibid.*, p. 223.
97. *Ibid.*, p. 223.
98. *Ibid.*, p. 215.
99. Michael Polanyi, *Personal Knowledge*, p. 299.
100. Thomas F. Torrance, *Space, Time and Resurrection*, Edinburgh, 1976, p. 20.
101. Thomas Langford, "Michael Polanyi and The Task of Theology", *Journal of Religion*, Chicago, vol. 46, January, 1966.
102. Karl Barth, *Church Dogmatics*, I. 2, Edinburgh, 1956, p. 535.
103. *Ibid.*, p. 540.
104. *Ibid.*, p. 544.
105. *Ibid.*, p. 581.
106. Thomas F. Torrance, *Space, Time and Resurrection*, p. 11.
107. *Ibid.*, p. 190.
108. Karl Barth, *Church Dogmatics*, I. 2, p. 545.
109. *Ibid.*, p. 580.
110. Michael Polanyi, *Personal Knowledge*, p. 123.
111. Michael Polanyi, *Science, Faith and Society*, p. 64.
112. Michael Polanyi, *The Study of Man*, p. 96.
113. Karl Barth, *Church Dogmatics*, I. 2, p. 587.
114. *Ibid.*, p. 589.
115. Michael Polanyi, *Science, Faith and Society*, p. 46.
116. Karl Barth, *Church Dogmatics*, I. 2, p. 590.
117. *Ibid.*, p. 590.

118. *Ibid.*, p. 588.
119. Thomas F. Torrance, *God and Rationality*, London, 1971, p. 201.
120. Karl Barth, *Church Dogmatics*, I. 2, p. 594.
121. *Ibid.*, p. 591.
122. Michael Polanyi, *Personal Knowledge*, pp. 156f.
123. *Ibid.*, p. 164.
124. Michael Polanyi, *Science, Faith and Society*, p. 45.
125. Karl Barth, *Church Dogmatics*, I. 2, p. 655.
126. *Ibid.*, p. 651.
127. Karl Barth, *Church Dogmatics*, I. 2, pp. 599–601.
128. *Ibid.*, p. 601.
129. Michael Polanyi, *Knowing and Being*, p. 56.
130. See p. 69.
131. See p. 39.
132. Michael Polanyi, *Knowing and Being*, p. 57.
133. Michael Polanyi, *Science, Faith and Society*, p. 59.
134. *Ibid.*, p. 63.
135. *Ibid.*, pp. 57–59.
136. Karl Barth, *Church Dogmatics*, I. 2, p. 628.
137. *Ibid.*, pp. 627f.
138. *Ibid.*, p. 655.
139. *Ibid.*, p. 655.
140. *Ibid.*, p. 649.
141. Michael Polanyi, *Personal Knowledge*, p. 322.
142. Michael Polanyi, *The Study of Man*, p. 67.
143. Michael Polanyi, *Personal Knowledge*, p. 322.
144. Karl Barth, *Church Dogmatics*, I. 2, pp. 709–710.
145. *Ibid.*, p. 711.
146. *Ibid.*, p. 716.
147. Michael Polanyi, *Knowing and Being*, p. 66.
148. Michael Polanyi, *Science, Faith and Society*, p. 64.
149. *Ibid.*, p. 50.
150. *Ibid.*, p. 51.
151. *Ibid.*, p. 83.
152. *Ibid.*, p. 52.
153. See p. 40.
154. Michael Polanyi, *Science, Faith and Society*, p. 57.
155. *Ibid.*, p. 59.
156. Karl Barth, *Church Dogmatics*, I. 2, p. 609.
157. Michael Polanyi, *Meaning*, p. 38.
158. Michael Polanyi, *Personal Knowledge*, p. 206.
159. Leslie H. Duncan, *Calling All Parents*, Edinburgh, 1980, p. 3.
160. *Ibid.*, p. 4.
161. *Ibid.*, p. 9.
162. Michael Polanyi, *Personal Knowledge*, pp. 279f.
163. *Decree on Ecumenism*, London, p. 28.
164. Rudolf Ehrlich, *Rome: Opponent or Partner*, London, 1965, p.

240. (Reprinted in *The Documents of Vatican II*, edited by W. M. Abbott, New York, 1966, pp. 34–366.)
165. *Ibid.*, p. 241.
166. *Ibid.*, pp. 241ff.
167. *Ibid.*, p. 245.
168. *Ibid.*, p. 245.
169. *Ibid.*, p. 245.
170. K. E. Skydsgaard, *One in Christ*, Philadelphia, 1957, p. 62.
171. Karl Barth, *Church Dogmatics*, I. 2, p. 548.
172. *Ibid.*, p. 549.
173. *Ibid.*, p. 549.
174. Georges Florovsky, *Bible, Church, Tradition, Collected Works*, Vol. I, Belmont, Massachusetts, 1972, pp. 75f.
175. *Ibid.*, p. 79.
176. Karl Barth, *Church Dogmatics*, I. 2, p. 549.
177. *Ibid.*, p. 551.
178. *Ibid.*, p. 546.
179. *Ibid.*, p. 566.
180. *Decree on Divine Revelation*, London, 1965, p. 20. (Reprinted in *The Documents of Vatican II*, edited by W. M. Abbott, New York, 1966, pp. 112–128.)
181. *Decree on Divine Revelation*, p. 21.
182. *Decree on Divine Revelation*, p. 11.
183. Georges Florovsky, *Bible, Church, Tradition, Collected Works*, Vol. I, p. 48.
184. Church of Scotland Panel on Doctrine, Edinburgh, 1977: *Agreement and Disagreement*, p. 5.
185. Michael Polanyi, *Meaning*, p. 186.
186. Thomas F. Torrance, *Royal Priesthood, Scottish Journal of Theology Occasional Papers* No. 3, Edinburgh, 1955, pp. 27–28.
187. See p. 88.
188. See p. 84.
189. Georges Florovsky, *Bible, Church, Tradition*, p. 79.
190. *Ibid.*, p. 80.
191. *Ibid.*, p. 92.
192. *Ibid.*, p. 101.
193. Thomas F. Torrance, "Truth and Authority", p. 219.
194. Michael Polanyi, *Personal Knowledge*, p. 265.
195. *Ibid.*, p. 9.
196. Thomas F. Torrance, *Theology in Reconciliation*, London, 1975, p. 270.
197. Rudolf Ehrlich, *Rome: Opponent or Partner*, pp. 254–255.
198. *Ibid.*, p. 255.
199. *Ibid.*, p. 256.
200. *Ibid.*, p. 256.
201. *Ibid.*, p. 265.
202. Georges Florovsky, *Bible, Church, Tradition*, p. 54.

203. Rudolf Ehrlich, *Rome: Opponent or Partner*, p. 268.
204. *Ibid.*, p. 268.
205. *Decree on Divine Revelation*, p. 12.
206. Rudolf Ehrlich, *Rome: Opponent or Partner*, p. 271.
207. *Ibid.*, p. 271.
208. *Ibid.*, p. 271.
209. K. E. Skydsgaard, *One in Christ*, p. 62.
210. Karl Barth, *Church Dogmatics*, I. 2, p. 627.
211. *Ibid.*, p. 628.
212. Rudolf Ehrlich, *Rome: Opponent or Partner*, p. 276.
213. Michael Polanyi, *Science, Faith and Society*, p. 58.
214. Michael Polanyi, *Knowing and Being*, p. 66.
215. Michael Polanyi, *Personal Knowledge*, p. 104.
216. Stanley L. Jaki, *The Relevance of Physics*, Chicago, 1966, p. 192.
217. Thomas F. Torrance, *Space, Time and Resurrection*, p. 174.
218. Georges Florovsky, *Bible, Church, Tradition*, p. 49.